The Ocean World of Jacques Cousteau

Guide to the Sea and Index

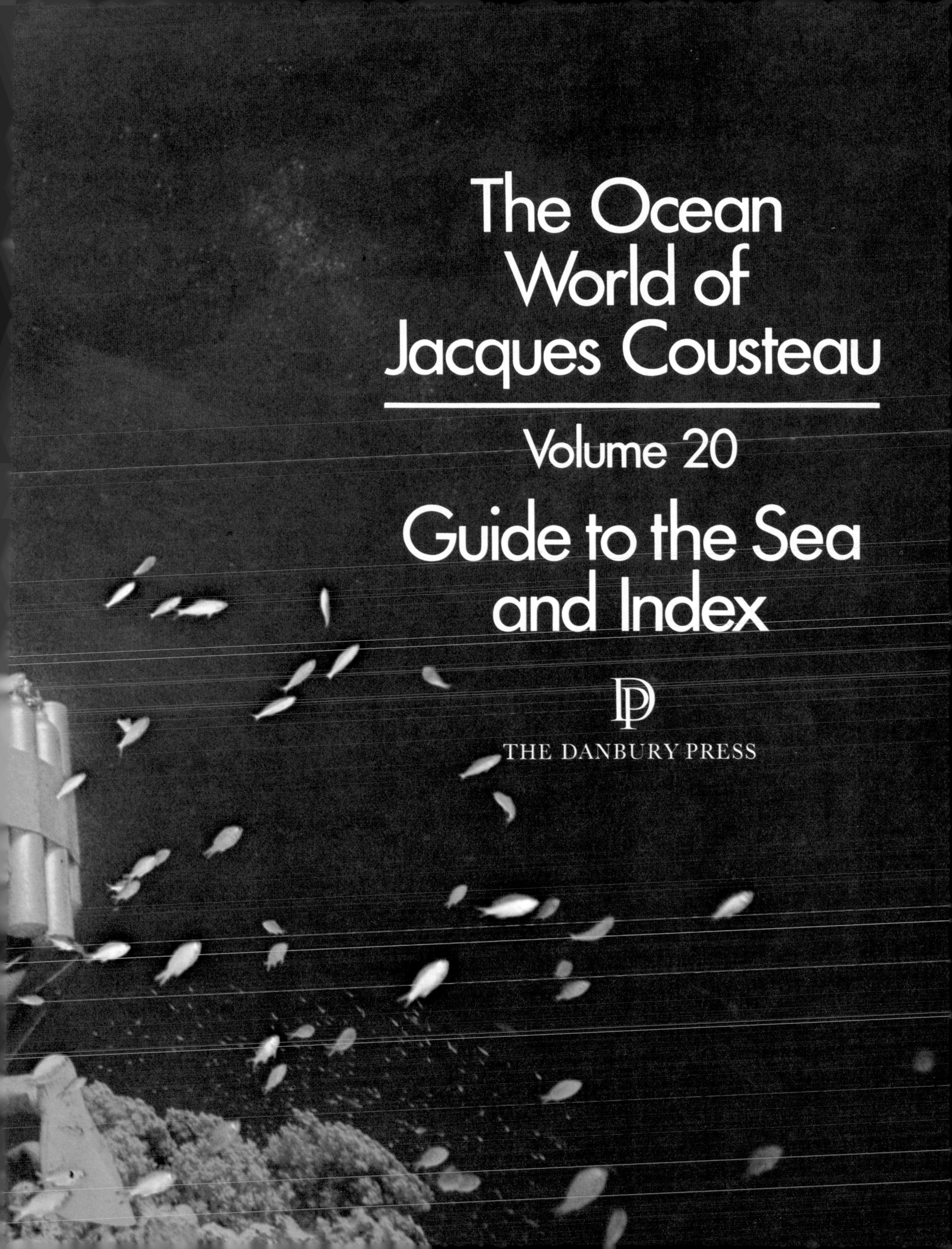

The Ocean World of Jacques Cousteau

Volume 20

Guide to the Sea and Index

THE DANBURY PRESS

Stranger than science fiction is this shrimp larva. With stalked eyes and delicate appendages, the planktonic creature will metamorphose into an ordinary bottom-dwelling mantis shrimp.

The Danbury Press
A Division of Grolier Enterprises Inc.

Publisher: Robert B. Clarke

Production Supervision: William Frampton

Published by The World Publishing Company

Published simultaneously in Canada
by Nelson, Foster & Scott Ltd.

ISBN 0-529-05166-4
Library of Congress catalog card number: 74-658

Printed in the United States of America
23456789987654

Project Director: Steven Schepp

Managing Editor: Richard C. Murphy

Assistant Managing Editor: Christine Names
Senior Editors: Ellen Boughn
Robert Schreiber
David Schwimmer
Editorial Assistant: Joanne Cozzi

Art Director and Designer: Gail Ash

Associate Designer: Terry McKee

Assistants to the Art Director: Martina Franz
Leonard S. Levine
Illustrations Editor: Howard Koslow

Vice President, Production: Paul Constantini

Creative Consultant: Milton Charles

Typography: Nu-Type Service, Inc.

Contents

It may be useful to draw a quick sketch of how our knowledge and understanding of the sea were obtained—how the SCIENCES OF THE SEA (Chapter I) have evolved from the earliest realization that there were vast mysteries to be unraveled by many different kinds of specialists. Today, marine scientists, regardless of their field, must be well schooled in a variety of areas in order to perform the tasks at hand. 8

The pursuit of knowledge is a process of gradual improvement and refinement in the perception of truth; but every once in a while comes one of THE TURNING POINTS (Chapter II), a revolutionary philosophical breakthrough. Each such turning point irrevocably alters man's relation with the sea and opens up new pathways for man's inquisitive mind to traverse, ever searching to broaden his horizons through new ideas. This list of breakthroughs, we hope, is not at its end. 20

We can deal with the sea in many ways, but only when we have experienced it from within, as a diver, can we know the thrill of being part of it. The beauty and serenity that the underwater world can hold for man is one of nature's truly marvelous gifts to us. We must remember, however, that we are not aquatic animals. Diving requires many precautions; we present an abridged guide to SAFE DIVING (Chapter III) to ease the way for the beginner's first venture into the underwater world. 36

How should we behave when we come to enjoy or to take advantage of the ocean? How can we best benefit from the sea without taking personal risks, but also without damaging an environment that is at the same time vital to us and so vulnerable? Some basic ADVICE FROM COUSTEAU (Chapter IV) may help us to understand, to fear, and to protect the sea and its creatures, some of whom are men. 46

Not all marine creatures are harmless objects of curiosity for the undersea explorer. There are many POISONOUS AND VENOMOUS MARINE ANIMALS (Chapter V) that must be avoided. Generally the inhabitants of the sea are not aggressive, nor do they lurk around waiting to victimize an unsuspecting diver. By understanding which groups are potentially harmful, anyone can safely enjoy the underwater world. It is worthwhile, though, to have an idea of what the best antidotes and treatments are. 54

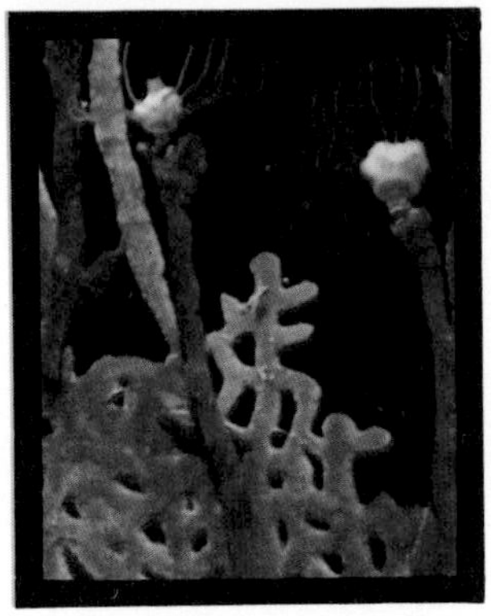

Throughout these volumes we have described and illustrated many marine organisms, each of which has a place in the classification used by scientists for purposes of communication. In this chapter on FILING SEA LIFE (TAXONOMY) (Chapter VI), we present a formal arrangement of all the plants and animals on earth, with the emphasis on life from the sea. 72

The units of TIME AND MEASURE (Chapter VII) are part of the everyday practice of oceanography. For reference purposes we include a metric conversion table, a geologic time scale, and a list of ocean data. They are becoming increasingly useful in everyday life. 88

Among the tens of thousands of pictures for these books were several that tell their own story—either of beauty or wonder or both. In a PHOTOGRAPHIC ESSAY OF THE SEA (Chapter VIII) we feature the best of these—and let them speak for themselves. The old adage "a picture is worth a thousand words" is never more vividly illustrated. 98

As long as human beings, directly as divers or indirectly through television, film, and photography, did not penetrate the sea, the ocean world could only be conjured from sketchy data. A WORLD WITHOUT WITNESS (Conclusion) hardly exists; but the diving witnesses attempted to tell their story throughout these books in the hope that it will help others understand, love, and protect the sea. 106

We have carefully avoided scientific terms all along in these books, but still there may be a number of words not familiar to nonscientists. The *Glossary* will attempt to define some of the less common terms used by scientists. The comprehensive definitions are, in themselves, a new source of information and can give the reader a valuable key. 108

For readers who have interests beyond the scope of our series or who wish to delve further into a particular subject, we have compiled a list of publications. In a short *Bibliography* we have subdivided the references that helped in the preparation of these volumes into a number of general categories. 122

In order that our readers might fully utilize all of the books in this series, we include in this volume a comprehensive *Index*. All subjects touched upon in these volumes are listed, as is characteristic of most indexes, in alphabetical order, giving the page and volume number on which the subject can be found, with photographs designated in bold typeface. 126

Chapter I. Sciences of the Sea

The study of geology encompasses all the subdisciplines of the earth sciences, and astronomy is the overall subject of space. Oceanology—or, as widely accepted, oceanography—is the science that deals with the sea. A practical way of approaching oceanography is to undertake the study of the specialty for which one has a particular interest—not necessarily with respect to the sea—and when this specialty is mastered, *then* learn about the sea in general and how one's area of study relates to it. This approach will help mold the student into both a specialist and an informed generalist, able to relate to many aspects of the science. For example, a detailed study of invertebrate ecology, both modern and paleoecology, will enable the biological oceanographer to understand the sea organisms of the present and past as well as provide background for future work on terrestrial animals. The idea is to develop options while concomitantly studying some area on a deeper level.

Specialization in the modern age has forever put an end to the person with general scientific knowledge—such as Copernicus, Newton, Galileo, Aristotle, and other great thinkers who contributed to the beginnings of the search for knowledge. When the general body of knowledge is small, every advance is magnified in proportion. Today many high school students superficially know as much as any of these learned figures, though few of them will have the time—or perhaps inclination—to understand what they are taught. Even fewer will later contribute to scientific advances.

Because of human nature, there will always be a division between pure and applied research. Applied research is devoted to a practical or commercial end. Certainly we could not do without it. But the very process of doing research with a point to prove prevents the researcher from wandering off on an unprofitable tack, and therefore a new idea might be lost. It is also part of human nature to prove your point, to find the answers. If the point is to make money or create a new product, truth sometimes is suppressed; the answers may be biased.

Pure research is simply the pursuit of information with no particular objective except perhaps to relate it to some other discoveries. One may think of pure science as a library full of carefully cataloged, but unsorted books. The applied scientist, as well as the industrialist, in need of detailed information about a particular subject need only know how to use the catalog; he will find the information, and he can trust it. Even if all unsound or unworkable new ideas are not winnowed out, the give-and-take of technical publications, seminars, and scientific meetings at least ensures the scientific community a chance to respond and rebut.

There are five general groups of oceanographers: chemists, physicists, marine geologists, marine biologists, and marine technologists. Universities today are stamping out graduates in these fields in astronomical numbers and an interested student should be absolutely certain that he is motivated by the excitement of knowledge, not by the prospect of lucrative jobs, because most new graduates are finding positions scarce. Unfortunately oceanography still has not made its importance felt with government and financial-granting institutions.

***Oceanographic research vessels** sometimes come in rather unusual shapes. FLIP (right) is a stable, drifting, sparbuoy-type platform.*

Narrowing the Field

The major branches of oceanography—physics, chemistry, biology, geology, and marine engineering—actually encompass a number of diverse but related subgroups.

The following constitutes a list of the subdisciplines contained within the five general fields of oceanography. Each is useful and necessary. Gathering all the information we can in each field is vital for a unified study of the environment. The word *marine* should be in the titles of most oceanographic disciplines but is left out to avoid repetition.

Physical Oceanography

Hydrology—the study of the physical properties of seawater; its role as a heat transfer agent

Hydrodynamics—water motion; diffusion and mixing of seawater

Optics—selective absorption, refraction, and scattering of light

Bathythermography—defining the frontiers of the various thermoclines

Acoustics—propagation, sound channels, and speed of sound underwater

Climatology—weather and seasons over the seas and their influence on landmasses

Meteorology—interacting of atmosphere and hydrosphere and resultant energies

Chemical Oceanography

Nuclear chemistry—the study of the diffusion, flocculation, and absorption of natural and artificial nucleids in nature and the laboratory

Physical chemistry—salts, nutrients, and particles in solution or suspension

Organic chemistry—compounds of biologic origins; molecular chemistry; clues to the origin of life

Salinometry—measurement of salts in solution

Equilibrium chemistry—reaction rates; dynamics of chemical reactions

Gas diffusion studies—action of gases in water; transfer through different density media

Biological Oceanography

Ichthyology—the study of cartilaginous and bony fish; their physiology and behavior

Mammalogy (Cetology)—whales, dolphins, sea lions; all sea mammals

Invertebrate zoology—animals without backbones; the physiology of marine animals as the basis for ecology and behavior studies

Botany and Planktonology—algae, phytoplankton; beginnings of food chains

Ecology—creatures considered as part of their surroundings; their behavior, interactions, reactions, instincts, communities

Pharmacology—medicines from the sea; how drugs act in the sea

Molecular biology—the intricate functions within a cell on an organelle level and on a chemical level.

Genetics—how life is transmitted

Population dynamics—productivity, interrelations, migrations, overpopulations, extinctions

Invertebrate paleontology—life of the past known through fossils

Vertebrate paleontology—fossil traces of past vertebrates: fish, reptiles, birds, mammals, and amphibians

Fisheries research—improving the catch; saving species from extinction

Mariculture research—controlled or artificial breeding of useful marine life

Marine Geology and Geophysics

Glaciology—the study of polar land and sea ice; the glacial periods

Vulcanology—undersea volcanoes; their dynamics, history, heat transfer through the sea bottom

Seismology—earthquakes, tremors, rifts

Bathymetry—measurement of sea depth; geological features of the sea bottom

Geomorphology—description of bathymetric features; their analysis and origin

Sedimentology—formation, deposition, and age of sedimentary deposits and rocks

Cartography—mapping

Mineralogy—chemical and physical aspects of rocks and sediment in and around the sea

Tectonics—mountain building, earth movement, continental drift; forces acting within the earth

Marine Engineering

Shipbuilding—the study of surface vessels, submarines, floating structures

Navigation research—improving direction and obstacle location

Power generation—solutions to the energy problem using the energy of the sea

Diving research—human and physical engineering to help man adapt to undersea life

Construction work—underwater construction, demoltion, tanks, jetties, landfill, dikes

Dredging—clearing channels and harbors

Mining—minerals, oil, salts from oceanic deposits on the shelf and abyss

Pipe and cable—laying, constructing, maintaining undersea lines

Photography, holography, cinematography—still, photomicrography, photomacrography, motion pictures, electron micrography

Oceanographers struggle with a ***bottom grab.*** *Its contents provide a glimpse of bottom conditions.*

Conshelf III in the Mediterranean

Nautical Milestones

1000 B.C.—Homer describes the earth as a flat disc encircling the Mediterranean. Hesiod refers to "lands beyond the sea," which he calls the Isles of the Blessed and the Hesperides.

600 B.C.—Phoenicians sail around Africa from east to west.

6th century B.C.—Miletus and Pythagoras, Greek mathematicians, declare the earth to be a sphere.

415 B.C.—Greek divers destroy booms at Syracuse.

4th century B.C.—Aristotle describes 180 species of marine life, showing how they can be grouped by body form.

330 B.C.—Alexander the Great is said to have been lowered into the sea in a barrel.

3rd century B.C.—Plato describes the civilization of Atlantis.

250 B.C.—Eratosthenes calculates the circumference of the earth at about 25,000 miles—very close to the fact.

1st century B.C.—Hippalus notes that the monsoon winds which tranverse the Indian Ocean blow one way half the year, then in reverse direction the other half.

Circa 100 B.C.—Poseidonius measures the depth of the sea near Sardinia at about 1000 fathoms.

1st century A.D.—Pliny the Elder, a Roman naturalist and writer, lists 176 species of marine animals.

2nd century A.D.—Ptolemy draws a map of the world, depicting the earth as a sphere.

196 A.D.—Divers cut ships' cables at siege of Byzantium.

1000—The Vikings sail from Scandinavia and reach Iceland, Greenland, and North America.

1250—Sir Roger Bacon, in *Novum organum,* describes air reservoirs for wreck divers.

1416—Prince Henry the Navigator of Portugal founds a school of navigation and the Great Age of Discovery is launched.

1472—Roberto Valturio of Venice builds a hand-propelled wooden submarine.

1487–1488—Bartholomeu Diaz sails around the Cape of Good Hope and back to Portugal.

1492—Christopher Columbus reaches the New World.

1497–1499—Vasco da Gama sails around the Cape of Good Hope, opening a new trade route to India.

Circa 1500—Leonardo da Vinci describes submarine boats and leather diving lungs with air hoses.

1511—Drawings of divers appear in print in an edition of Vegetius' *De re Militari.*

1519–1521—Ferdinand Magellan sails to the Pacific and is killed in the Philippines. Juan del Cano completes first circumnavigation of the globe.

1535—Diving bell is used to explore sunken Roman galleys in Lake Nemi, Italy.

1537–1590—Gerhardus Mercator, a Flemish geographer, creates a new kind of map still in use today.

1576, 1577, 1578—Martin Frobisher, an English explorer, seeks a northwest passage from Europe to India.

1578—William Bourne, an Englishman, builds a submarine with ballast tanks operated by jack screws.

1585, 1586, 1587—Englishman John Davis makes three attempts to discover a northwest passage.

Catamaran vessel of the South Pacific.

Bottom grab retrieves a sample.

Cachalot diving chamber.

Troika, a deep-sea camera sled.

1609–1612—Henry Hudson, an English explorer, discovers the Hudson River and Hudson Bay.

1616—William Baffin discovers Baffin Bay and is convinced there is no ice-free northwest passage.

1620—Cornelius van Drebbel builds a submarine with a primitive air-purifying system and tests it on the Thames.

1640—Jean Barrié, a Frenchman, uses a diving bell in a wreck.

1642—Abel Janzoon Tasman, a Dutchman, sails around Australia. Tasmania is named after him.

1643—Cossack 40-man cowhide submersibles attack Turkish ships in the Black Sea.

1644—Père Marin Mersenne, a Frenchman, designs a fish-shaped metal submarine with snorkels.

1660—Robert Boyle discovers a law governing the physical properties of compressed gases.

1679—G. A. Borelli, an Italian physicist, designs a self-contained diving dress.

1690—Edmund Halley's diving bell reaches a depth of 60 feet and stays submerged for 90 minutes.

1692—Denis Papin, the French inventor of steam engines, develops a submersible chamber.

1715—John Lethbridge, an Englishman, uses a leather diving suit to 60 feet on salvage.

1725—Count Luigi Ferninando Marsigli of Italy publishes *The Physical History of the Sea.*

1729—Nathaniel Symons's submarine boat dives for 45 minutes in the River Dart, England.

1736—Carolus Linnaeus, a Swede, introduces a system for classification of plants and animals still in use today.

1754—Richard Pococke describes a helmet suit with pumped air used on salvage at Yarmouth, England.

1768–1779—James Cook, on the first scientific ocean expeditions, makes soundings to depths of 200 feet.

1770—Benjamin Franklin's map of the Gulf Stream is published and becomes a great aid to shipping.

1802–1804—Alexander von Humboldt describes the Pacific current off South America that bears his name.

1807—French physicist Arago advocates echo sounding.

1818—Sir John Ross collects worms and starfish from a depth of 1050 feet, proving that life exists at great depths.

1831–1836—Charles Darwin on the *Beagle* develops his theory of the formation of atolls.

1837—Augustus Siebe's "closed" diving dress is introduced, the ancestor of today's helmet suit.

1839–1842—Charles Wilkes, an American, uses copper wire to make soundings.

1855—Publication of *The Physical Geography of the Sea* by Matthew Maury, a U.S. Navy officer.

1863—*Plongeur,* first proven powered submarine, is launched.

1864—Svend Foyn, a Norwegian, invents the explosive harpoon, marking the beginning of modern whaling.

1868–1870—Wyville Thomson and W. B. Carpenter lead expeditions on the British ships *Lightning* and *Porcupine.*

1869—Publication of Jules Verne's *Twenty Thousand Leagues Under the Sea.*

1872—The first marine biological station is founded at Naples, Italy, by Anton Dohrn.

1872–1876—Scientific oceanography, developing every branch of the study, begins with the British *Challenger*'s round-the-world expedition under Wyville Thomson.

1873—Publication of Wyville Thomson's *The Depths of the Sea,* a compilation of data from four scientific cruises.

1873—Louis Agassiz founds first marine biological station in the U.S., near Cape Cod.

1877–1905—Alexander Agassiz conducts important voyages on the *Blake* in the Caribbean and Gulf of Mexico, and on the *Albatross,* the first ship specifically designed for oceanographic research, in the same areas and in the Pacific.

1879—The laboratory of the Marine Biological Association is founded at Plymouth, England.

1882—Studies in patterns of fish migrations are undertaken by the Fishery Board for Scotland.

1886–1889—The Russian ship *Vityaz* collects pioneer data on sea temperature and density around the world.

Deepquest, **a research submersible.**

Calypso **diver with modern propulsion unit.**

1886–1906—Prince Albert I of Monaco outfits four yachts as oceanographic research ships and does extensive work in the Atlantic.

1889—The German "plankton expedition," under the leadership of Victor Hensen, makes the first quantitative investigations of the biological productivity of the seas.

Diving archaeologists work a wreck.

The U.S. Navy's *Sealab III.*

1893—Louis Boutan, a Frenchman, takes the first underwater photographs at Banyuls-sur-Mer.

1893–1896—The Norwegian Fridtjof Nansen allows his ship the *Fram* to be frozen in the arctic ice in order to study the polar icecap.

1894—C. G. J. Peterson, a Dane, develops a theory of overfishing and invents a fish tag.

1899—Oceanographic Institute of Monaco is built.

1900—French physicist d'Arsonval proposes to the Academie des Sciences a power source from small differences in sea water temperature.

1902—The International Council for the Exploration of the Sea is founded.

1905—Scripps Institution of Oceanography is founded.

1906—British Admirality Deep Diving Committee, under John Scott Haldane formulates decompression tables.

1909—Admiral Robert E. Peary is the first man to reach the North Pole.

1910—The Intergovernmental Commission for the Scientific Exploration of the Mediterranean Sea (CIESMM) is created in Monaco.

1910—Johan Hjort, a Norwegian, leads the *Michael Sars* expedition in the North Atlantic.

1911—Roald Amundsen reaches the South Pole.

1912—Publication of *The Depths of the Ocean* by Sir John Murray and Johan Hjort.

1917—Paul Langevin develops a high-frequency sound generator applicable to underwater echo sounding.

1915—Alfred Wegener presents his theory of continental drift.

1920–1922, 1928–1930—Cruises of the *Dana,* one result of which is discovery of the life history of the eel.

1925—British scientists aboard *Discoverer* study antarctic whales and their environment.

1925–1927—The German *Meteor* expedition discovers sea mounts in the Atlantic.

1928-1929—George Claude demonstrates a 60-kw power plant near Liège, Belgium, utilizing temperature differences between ocean surface and bottom waters.

1930—Woods Hole Oceanographic Institution is founded.

1934—William Beebe and Otis Barton descend to 3036 feet off Bermuda in the bathysphere.

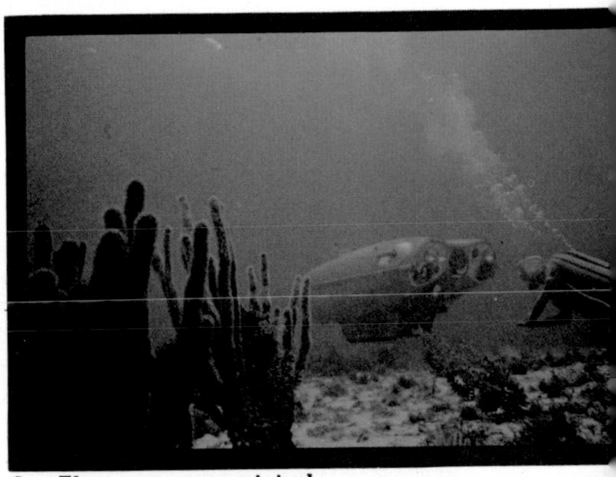

Sea-Flea, a one-man minisub.

1942—Sverdrup, Johnson, and Fleming publish their comprehensive study entitled *The Oceans.*

1943—Fully automatic compressed-air aqualung is invented by J.-Y. Cousteau and Emile Gagnan.

1943—Marine Laboratory of Miami is founded.

1945—Börje Kullenberg, a Swede, develops a piston corer that takes sediment samples 70 feet long.

1947—Harold Urey dates ocean sediments by measuring the ratios of oxygen isotopes.

1947–1948—Round-the-world cruise of the Swedish *Albatross* under Hans Pettersson emphasizes marine geology and dredge samples are taken from 26,500 feet.

1950–1951—The Danish research vessel *Galathea* trawls fish from 23,000 feet and other animals from 33,000 feet.

1950–1952—The Challenger Deep in the Mariana Trench is sounded at 35,847 feet, the deepest in the ocean.

Star II, undersea observation submersible.

1952—E. Steeman-Nielsen on the *Galathea* uses carbon-14 techniques to measure primary productivity of the sea.

1955—The first atomic submarine, *Nautilus,* is launched.

1957—J. C. Swallow's neutrally buoyant float aids the discovery of a south-drifting current beneath the Gulf Stream.

1957–1958—International Geophysical Year emphasizes cooperative oceanographic studies.

1958—First United Nations Conference on the Law of the Sea meets in Geneva.

1958—U.S. nuclear submarines *Nautilus* and *Skate* sail under the North Pole. *Skate* surfaces near the pole.

Diver returns to *Conshelf II.*

Conshelf III diver with umbilical support

1959—The first meeting of the International Oceanographic Congress is held in New York.

1959—The first exploration submarine, Cousteau's diving saucer, is launched.

1959–1965—International Indian Ocean Expedition groups ships and scientists of 22 nations.

1960—Creation of UNESCO's Intergovernmental Oceanographic Commission (IOC).

1960—Second United Nations Conference on the Law of the Sea meets in Geneva.

1960—Jacques Piccard and Donald Walsh in the bathyscaphe *Trieste* conquer the Challenger Deep, 35,800 feet down.

1962—Launching of Cousteau's *L'Ile Mystérieuse,* the first manned stable platform for ocean research and specially designed to be anchored in the deep sea.

1962—FLIP (Floating Instrument Platform) launched by Scripps Institute in California.

1962—Hannes Keller swims briefly at 1000 feet after being lowered in a chamber off coast of California.

1962–1963—Conshelf I and II are first underwater habitats.

1963–1964—International Cooperative Investigation of the Tropical Atlantic by seven nations begun.

1964—U.S. Navy Sealab I, at a depth of 193 feet, is occupied by four men for eleven days.

1964–1970—*L'Ile Mystérieuse* anchored in 8600 feet of water in Mediterranean as a base for scientists.

1965—Sealab II, U.S. habitat 205 feet deep off California, and Conshelf III, French habitat 328 feet deep in Mediterranean, are connected by telephone.

1965–1966—Multinational cooperative study of the Kuroshio Current and adjacent regions of the Pacific.

1966—First full-size electric generating plant uses tidal energy at the Rance River in France.

1968—*Glomar Challenger* begins open-ended expedition drilling in the deep sea.

1972—U.S. Navy divers W. Ramos and C. Delucchi work in sea for 30 minutes at 288 meters (950 ft.) off San Clemente.

1972—In Marseilles, France, COMEX divers R. Gauret and P. Chemin stay 80 minutes at a simulated pressure corresponding to 610 meters (2013 feet).

1973—Earth Resources Technology Satellite launched to survey productivity and pollution of the seas.

1973—Inaugural session of the Third United Nations Conference on the Law of the Sea opens in New York.

1973–1974—Franco-American geophysical and diving operation FAMOUS in the Rift Valley of the Atlantic Ridge.

1974—Second session of the Third United Nations Conference on the Law of the Sea meets in Caracas, Venezuela.

Conshelf III and tendership.

Chapter II. The Turning Points

The idea of wanting to understand something about the nature of the sea was the first philosophical breakthrough in marine science. Long before cave painting was in vogue, a man must have wondered about the sea. He and his ancestors had probably fished, swam, and collected shellfish in and from the sea, but this was different. Certainly the identity of the first marine scientist will never be known—there were probably many who independently were thinking about the same questions—but just by wondering, he deserves to be listed ahead of those who later revolutionized the science of the sea.

Progress has been the result both of a continuous improvement in technique or technology and of rare discontinuities brought about by bold innovations; such a process is not unlike the slow adaptation and abrupt mutations in the evolution of life. For example, the introduction of a compass in navigation was an important breakthrough because it related the routes of ships to a geographical system of reference even when the ship was out of sight of land, but the many improvements made over the centuries to compasses, including the gyrocompass, were only part of the general advances made in technique.

We are here only concerned with the single concepts that led to entirely new approaches in the use, the understanding, the study of the sea. Each concept we consider a turning point in man's relationship with the sea opened a new dimension to the body of knowledge: dimension in the strict sense here means a new direction of expansion, not a mere addition. The first philosophical breakthroughs were the development of the hull, the sail, and the steering oar. Throughout history later important turning points include the compass, astronomical navigation, maps, remote eyes (sonar and radar), diving, cutting the ties with the surface, the ocean as part of a global thermodynamic machine, evolution, mariculture, the study of the sea from outer space, and now a global concept of the law of the sea. This chapter will consist of a brief discussion of each of these so-called philosophical breakthroughs, especially their relationship to a general view of marine science.

In the beautiful, clear waters of ***the coral sea*** *(left), early man gathered food, swam for pleasure and probably began to wonder about the sea. Man has changed since then, but the coral sea, although endangered, remains to inspire man to wonder.*

Trails of bubbles (right) mark the course of Calypso *divers who must light their path artificially at depths* ***where the sun fails.***

The Hull

The principle of buoyancy is so obvious to us today that it is difficult to conceive of a time when it had not occurred to any human to think of riding on a float. The breakthrough, in essence, was the realization that things that float could be ridden.

The principle of buoyancy is merely a side effect of the weight difference between water and a floating substance. When the amount of displaced water is greater than the weight of an intruding object, the object floats. So many things are buoyant that it is difficult to define which natural object made up the hull of the first boat. Driftwood, bark, reeds, a large dead animal, or more probably a fallen tree trunk was the earliest hull, or at least was the model for it. The developments in shipbuilding after the discovery of simple buoyancy were evolutionary: skin, basket, and lashed raft boats probably came next, followed by the beautiful and graceful hulls in use by the Phoenicians, Greeks, Polynesians, Norsemen, and others of earlier times. The progress of shipbuilding appears to have been quite rapid once the basic notion was implanted in many minds.

Man has crafted boats from the materials his culture and environment provided. Where baskets of ***woven reeds*** *carried goods, they also served to carry men on the water (below), and where woodcraft was practiced,* ***boats and rafts of logs*** *crossed rivers and seas (right). As these contemporary photographs indicate, such primitive boats are still fashioned and used today.*

Australian aborigines still often use primitive rafts. The voyages of *La Balsa* and *Kon-Tiki* proved the basic seaworthiness of the concept. The stretched-skin boat is in constant use in the arctic and also in Asia. The bamboo, reed, and papyrus boat, as well as the woven, caulked basket, serve just about

every continent somewhere. There is no reason for the modern yachtsman to sneer at the primitive-hulled boats since they serve in a way that his expensive, polluting toy does not: they can be created on the spot when they are needed, and they can be dismantled or discarded afterwards.

The general shape of most hulls has remained the same since early historic times. This has occurred partly because of streamlining, partly for control—topics of the next breakthrough, controlled motion. But if the shape of hulls appears to have remained constant, the modern variations on hull materials run the gamut of natural and synthetic substances: wood, metals, plastics, cement, canvas, and rubber. There are double hulls, icebreaking hulls, racing hulls, unsinkable designs. The breakthrough is not the design but in the simple concept of having a support that floats in water—a prehistoric turning point in man's entry into the sea.

Modern Arab dhows (above) are rigged as simply as were the boats of a thousand years ago.

Controlled Motion

The breakthrough of controlled power and steering control happened long before recorded history. The ocean currents were probably used as the first medium of propulsion when they happened to go exactly where needed. Oceanic currents, for example, helped the trade winds to pull the *Kon-Tiki* across the Pacific; but the raft was equipped with sails, tiller, movable centerboards, and a pointed bow. For any serious venture in sea travel it was necessary to have means to propel and control a vessel at the whim of the sailors.

The shape of a boat's hull is in part a controlling device; it is, for example, the centerboard on a small sailing vessel that prevents side-slip and allows tacking and most maneuvers. The earliest ships were probably poled through shallow water; these were followed by paddled boats, rowed boats, (after the invention of the tholepin), and then someone put together the set of ideas that produced the sail. Because the oldest extant ship pictures show pointed bows, we realize streamlining occurred very early. In spite of this, some recent boats (such as round American Indian bullboats) are conspicuously unstreamlined.

Whatever was the sequence of invention, the concept of controlling motion while afloat was the turning point in practical water travel. Refinements such as motors, hydrofoils, outriggers, catamarans, as well as the ancient additions of sails and paddles, are simply logical extensions of the central concept of controlled motion at sea.

Open-Sea Navigation

The term *to navigate* has several meanings, two of which are "to steer" and "to travel" (usually by ship). Navigation began with dead reckoning: direction by sightings of land. This technique was practical when visibility was adequate and the boat stayed close to shore. If a course was plotted on the basis of a single point, current and wind drifts could run the ship far astray. It is in the sense of steering out of sight of land, in the open ocean, that navigation was a significant turning point in ocean sciences. It was a two-stage development: using the stars and using the magnetic field of the earth with a compass. Correlating the positions of the stars with the location of the observer involved an elaborate understanding of cosmology. At first the North Star (the polestar) was used as a simple indication of direction North. The principle of triangulating bearings followed the inventions of devices for accurately measuring distances on the ocean or land surface, and angles with respect to celestial objects. When observations are made with two or more objects, the position of a ship can be continuously determined by the change in the angular relationship with the observed points. The quality of the technique depends on the accuracy of the angular measurements, and the knowledge of their true positions. When the early astrolabe and sextant later made their appearances, measuring the angle from the polestar to the observer and to the horizon gave a quick and accurate measurement of latitude. But the calculation of longitude needed an accurate knowledge of time, and it was only possible with the invention of chronometers. The accuracy of ships' chronometers could mean life or death, and the maintenance of this precious clock was the captain's privilege. For centuries with the aid of the sextant and the chronometer ships determined their position, within one mile, by measuring the angles of two or more stars or of the sun at two different times. The little magnetized needle called the compass, a Chinese concept, allowed ships to steer accurately between two sets of astronomical measurements. Modern improvements such as the satellite or inertial navigation are simply evolutionary technological achievements. Knowing where you are at sea is the real turning point.

Fast sailing ships *(below) reached their commercial peak in the late 19th century. Today they may be revived because they do not consume petroleum.*

Charting the Earth

There are philosophers, often called empiricists, who believe that nothing exists unless someone is there to observe it. Thus a forest has no bird songs if a deaf person walks through it. This view is highly anthropomorphic, but it helps to illustrate a point: without maps, there is virtually no such thing as discovery and exploration because subsequent use and knowledge of the explored areas cannot be obtained. As one of the philosophical breakthroughs in matters pertaining to the sea, the map cannot be underestimated. Maps have been to navigation what writing has been to civilization.

Early maps date back to the period of Greek and Phoenician navigation in the first millennium B.C. They avoided one of the basic obstacles cartographers must deal with today. The earth as an approximate sphere does not lend itself to represention on a flat map. In spite of concepts of a spherical earth in antiquity, ideas about the earth before Columbus's time featured a flat earth, and the problem of projections didn't interfere with the mapping game.

When European civilizations rediscovered the spherical earth, the process of projecting maps on flat pages took on new meanings. Latitude and longitude grids have been placed on maps since the days of Ptolemy (150 A.D.). Various systems of projections have been used—the Mercator, gnomonic, and polar projections, and many variations on the equal-area concept. No perfect multipurpose map has ever been developed, but the Mercator system, projecting the features from the center of the earth onto a cylinder parallel to the lines of the poles, has been universally adopted for practical navigation maps. Maps have served to lead the great discoverers of the sea to the known so that exploration of the unknown could follow.

*The **electronic eye** (above) pierces distances and obstacles in a way no human or hawk eye can match.*

To See Beyond the Eye

Up to this point in the discussion of the turning points of marine science and history, we have been cataloging the fundamental events that allowed mariners to sail and chart the world's oceans. The order has been approximately evolutionary because they follow in logical sequence: the hull had to precede the control of motion; both had to come before navigation and maps. Each of these concepts revolutionized the relationship between man and the sea. Sailors could travel anywhere by sea—as long as conditions were favorable. The remainder of the breakthroughs are those which allowed hidden facets and potentials of the ocean's realm to be explored.

To see beyond the limits of the eye had been a dream of mankind for centuries. The first telescopes turned a part of the idea into reality, but seeing beyond the eye really meant seeing without the eye. This dream had only to be aimed at the impenetrable depths or at the thickest fog to generate the concepts of sonar and radar.

Sonar makes use of the excellent sound-transmitting ability of water: by sending out pulses and recording the delay of return, the distance and direction to a foreign object underwater can be determined. This object can also be the sea bottom; modern side-looking sonar can depict a deep area almost as well as an aerial photo can.

Radar uses very short radio waves above water to give the distance and direction of all surrounding objects, even in darkest night. In fog, rain, or snow, radar can give warnings of the presence of icebergs and of other vessels. The mechanics involved in electronic navigation are beside the point. The philosophical breakthrough occurred when man decided that he would see without his eyes, through water, through the sediments of the sea bottom, as well as through the clouds and the long polar nights.

Getting In

Since antiquity men probably employed free diving—and probably much of this diving was for food gathering; but it is difficult today to measure the amount of boldness, of sheer courage, that was necessary to make the first plunge. When man deduced that he could extend the time of his visit in the sea, the breakthroughs occurred. The ocean was a frightening and treacherous element. The need for food gathering was probably less of a motivation than man's deep-rooted instinct for challenge.

Once the decision was made, primitive man faced extraordinary limitations—underwater he was blind, cold, and deprived of air; over the centuries he would have to use all of his ingenuity to correct these liabilities. We have reason to believe that our remote ancestors were as successful in their diving as the Greek sponge divers of 1700—two and a half minutes as far down as 250 feet. From this point to modern times, many precious improvements were achieved in stages. First, small goggles were made from pieces of broken glass inserted into hand-carved coconut cups. Then diving stones and baskets were used; later the diving bell, the helmet, wetsuits, and submersibles were employed.

Progress was slow at first, mainly due to lagging knowledge and technique rather than a lack of imagination. Leonardo da Vinci conceived most of today's gear—but it was merely theoretical, for none of his inventions could have been built at the time. However, man did "get into" the sea when he realized that the water was not an enemy to be invaded but rather a home that could be visited and lived in when the proper furnishings and amenities had been acquired.

Helmet dress *is still used in special conditions, but most divers today are free of the surface.*

Cutting the Ties

Going into the sea has revealed that the element that had frightened primitive people indeed holds hundreds of mysterious traps. Some examples are caisson disease or bends; rapture of the deep, also known as nitrogen narcosis; and air embolism. There were many other dangers, especially when the long lines and pipes and cables from the surface got entangled due to currents—these lines were frequently cut when jagged shipwrecks were entered, often severing the diver's air supply. It was a breakthrough when it was realized that these "life lines" were really death lines; actual safety was in freedom.

The concept of bathyscaphe rather than bathysphere and of aqualung rather than helmet dress opened a new era of discovery. With its demand valve the aqualung was the first practical approach to liberation in the sea, and the sea at last became a temporary home to get into. The freedom to explore undersea caves, to swim long distances underwater, even to attempt to live in the sea only came about when the ties binding divers to the surface were cut.

Vessels too had been independently modified to cut the ties with the surface. The modern nuclear and research submarines can perform extraordinary feats of diving depth and duration. The crew members of some nuclear subs can almost be classified as true undersea creatures—except that they are encased in an artificial vehicle.

We will finally turn the conceptual corner when man is able to live and remain in the sea. Conshelf and other experiments have proven the feasibility of long tenure below the surface, but that is visiting for science and not for living.

The submersible allows the undersea scientist almost free rein to ***travel anywhere underwater.***

The Thermodynamic Ocean

Ancient sailors knew about the tides and about winds—such as the trades or the westerlies. They knew about the large seasonal fluctuations and about the general behavior of storms. Most of their knowledge was empirical—observation in the course of their lives on the sea. The breakthrough in oceanographic research came when physical oceanographers began to think of the sea as a thermodynamic machine—with masses of water moving in systematic layers, and with storms, waves, currents all consequences of heat exchanges between the sun, the sea, the atmosphere, and the rotation of the earth.

When a substance is at a temperature above absolute zero (0° Kelvin, −459.72° F.), it has heat, and its molecules are in motion. There is a vast storehouse of heat in the ocean (which is never much colder than 32° F.), and this heat comes from the sun and from deep in the earth's core. The difference in temperature between surface and deep waters is, as we have seen in Volume XVII, *Riches from the Sea,* one of the most promising sources of energy on this planet.

Water in the sea is stratified by density, and density differences are caused by temperature and salinity gradients. The continuous-reading thermometer and salinometer were the developments needed to map these water masses. With this set of innovations came an understanding of the deep-water circulations and resulting enrichments in nutrients which supply the primary producers.

The surface currents and winds can be charted and studied with insight, now that we understand such things as the Coriolus effect and the air-water temperature interactions. It was the realization that we are surrounded by oceans which are simple heat engines that began all of these studies.

Sea-Floor Spreading

A revolution in the science of marine geology is happening right now. Old ideas concerning the formation of the sea floor and the long-unanswered questions about the deep sedimentary deposits are being rethought and reanswered. The revolution began—in the sense that it began to have effect—in the 1960s; but thoughtful men had recurrently proposed the basics of it since the Middle Ages. Scientists now must consider the

spreading of the sea floor and the consequent continental movement.

The question of why the sea bottom sediments are no older than the Mesozoic era is now easy to answer. They were swallowed up once again by the earth. We are beginning to see that ocean basins came about due to sea bottom motion, and we can study them from the perspective of their earlier positions. We know why trenches are near volcanic island areas, and we can almost predict earthquakes and volcanic eruptions.

There has been a continuous controversy within the geological establishment since the evidences for continental drift became overwhelming. The venerable, but still strongly doubted theory was best proven by sea-floor spreading. The motion of the sea floor, at about 1 to 5 cm per year, is slow to human perception, but fast enough to create thousand-mile rifts over geologic periods. In acceptance of the new global tectonics, old theories on mountain building and prehistoric events must be rejected.

Evolution

We have used the word *evolution* throughout this chapter to describe the gradual changes in technology and thinking which follow the new discoveries in marine science. Now we must deal with the most-often-used concept that goes with this word: the evolution of living creatures; that is, organic evolution. What is important is to assess how much the idea of organic evolution has changed man's perspective of the world around him, and how he presently views the sea. When it was thought that all life began in the garden of Eden, it was heretical to conceive of the ocean as being older than the third day of creation—4004 B.C. as defined by the Bishop of Ussher in 1654. The thought of humans and apes evolving from a common ancestor, inherent in the theory of evolution, was especially offensive to strict interpreters of biblical genesis. After the battles of the Darwinians against the theologians had ended, scientists had a more realistic view of the time period within which marine phenomena occurred—about 4 billion years by recent estimates. The explanations of many of the long-term marine developments (such as erosion, deposition, fossilization, and finally sea-floor spreading) can now be evaluated in their proper perspective.

The many and varied sea creatures lost some of their "monstrousness" when they were seen from the viewpoint of evolution. The strange appearance of, for example, the deep-sea anglerfish is logical and a functional answer to its problems of living in very special conditions; it could be viewed as such after the Darwinian survival of the fittest became a standard precept of the biologist.

The divergent ***pathways of evolution*** *are shown by the Galápagos land (below) and sea iguanas.*

Commercial reptilian meat is raised, in much the same way as beef, on a ***modern turtle farm.***

Mariculture

To reap you must sow—at least on land, and with domesticated crops. Before the idea of cultivating plants occurred to someone in prehistoric times, the harvest was obtained without effort from wild grasses and fruits. The diet was supplemented by meat from hunting; and the populations, always small in number, were seminomadic, moving when an area became too depleted to provide enough food. The limitations are obvious: only a small localized population can be supported, and considerable energy must be expended to obtain the food. The first agrarian settlements were only possible when humans had accumulated the knowledge necessary to understand which plants could be cultivated, which animals could be efficiently raised. Then the hunters became farmers, and the human race proliferated.

What man and science had done extensively with land plants and animals has been only touched upon in the sea. In certain fields though—for example, oyster or mussel farming—mariculture is an ancient practice of the Orient and fairly old in Europe. But high efficiency, rational farming of popularly accepted protein-producing animals is now generally experimental and on a small scale. The real breakthrough in mariculture is yet to come. It is difficult to imagine why we do not devote a major portion of our funding toward developing and using mariculture to meet our needs. All this time, fish and whales are exploited to extinction, shellfish are being polluted, overfishing is rampant as mankind waits for its leaders to have their long overdue moment of enlightenment.

The Sea from Space

The launching of several weather satellites, as well as a series of Earth Resources Technology Satellites (ERTS) launched in 1972 and the first manned orbiting space station, Skylab I, launched in 1973, initiating a pragmatic breakthrough in monitoring the aspects of the oceans. Some scientists have maintained that we are approaching the study of the sea from the limited perspective of "microtools for a magna ocean-world." With the use of the synoptic view of the satellite, a true magnatool is here at last.

Although the ERTS satellite is only a year old, data that has come from it is already in use, and one of *Calypso*'s functions in its recent antarctic voyage was to correlate measurements made from an orbiting satellite with on-site data taken by a surface vessel (for temperature and photosynthetic production in seawater). When all the satellite meters are calibrated and when a set of instrumented buoys are emplaced in position around the world's oceans, the routine data collection from space will free research vessels to explore the more diverse and novel problems of oceanography.

The weather satellites (such as NIMBUS) have proven themselves many times over—as anyone who watches television weather reports must realize. The satellite photos show cloud covers and patterns of motion for air masses as well as the limits of polar ice. The view of the sea from space as a result of the use of satellites is a significant turning point in marine research. Someday we will have a view of the entire ocean at once, with many bits of data simultaneously recorded worldwide. If we can quickly survey the entire ocean, perhaps we can save it.

During its orbit around the earth, ***the eye of the satellite*** *homes in on the Gulf of Mexico.*

The Global Ocean

At the moment we have a continuous stream of horrors and absurdities in our daily affairs with the sea. International squabbles about fishing rights and national limits are preventing the safe and sane use of the still-bountiful resources in the sea. International accord on fishing and whaling is still thwarted by the greed of the few nations that have the largest investment in their fishing fleets.

The overall primary productivity figures for the sea indicate that there could be enough living energy available to nourish human population to any limit it might reasonably expand to. When all peoples of the world finally cooperate in affairs of the sea, then a proper share of this resource can be apportioned. By the nature of the movement of water masses in the ocean, the areas of highest productivity are concentrated in certain localities, such as around the antarctic continent. Until the sea is for everyone, no nation will, nor can afford to, devote the proper amount of effort to developing methods to manage such resources properly for fear that the countries that are fortunately located in productive areas (such as Peru on the Humboldt Current) will claim proprietary rights to the end products.

The earth and its occupants could be heading for trouble. We think we have the opportunity to achieve the golden age of the global earth and ocean, but always imminent looms a black age of destruction, or almost-total destruction. At the moment this destruction looks inevitable. It may come from the atomic blast; from the destruction of marine plant-life through herbicides, chemicals, heavy metals or the lowering of light penetration due to air pollution; or it may come from directly poisoning the sea until it cannot support life. Many effects may come together, reinforce each other (synergesis), and condemn the sea. Until we pool our knowledge we will not know what the other is adding or removing to and from the sea. If the golden age comes about, the sea will belong to everyone. When man shares the sea with his neighbors and the sea-life, he will have turned the corner of one of his blindest alleys. No one should have ownership of a seashore; there is no such thing as the right to a private beach. The areas of shore are too small with respect to the size of human populations and landmasses to be restricted by the whims of the fortunate. The global ocean includes the free shoreline and in purest form the unpoisoned rivers and streams that flow into it. This breakthrough is unfortunately not going to come easily.

Our planet seems boundless in this view from space, but it may be ***doomed unless we cooperate.***

Chapter III. Safe Diving

The sudden boom in sports diving caught legislators without regulations to govern the new men-fish. There were no licenses needed for equipment, dealers, or divers. What fishing laws could apply to persons who captured their prey out of sight of the authorities? What steps could be taken to assure that a diver's equipment and breathing gases were safe? What treasure and salvage rights applied to divers? Who was in a position to police the underwatermen?

These problems have been met in various places with differing solutions; diving laws vary sharply along arbitrary lines. The international rules for divers are as inconclusive as those dealing generally with the use of the seas, and reflect a lack of planning. In many countries, Great Britain and the Soviet Union, for example, diving organizations are responsible for controlling underwater activities. They have set standards for equipment and have formulated training programs. They have agreed on diving etiquette to prevent restrictive legislation.

Diving in the United States is also largely self-regulated. Safety depends on thorough training, good sense, and ever-improving equipment. Four national organizations have agreed upon standards for diver certification. They are the Professional Association of Diving Instructors (PADI), National Association of Underwater Instructors (NAUI), YMCA, and National Association of Skin Diving Schools (NASDS). Many compressed-air dealers will fill only tanks to be used by C-card (certified) divers. The tanks themselves must meet Department of Transportation specifications to bear a DOT imprint, and must be checked every five years and marked with maximum safe capacity. The effectiveness of self-regulation is evident in diving's low casualty rate.

There are two aspects of diving that are subject to legal regulation: the taking of live animals and the recovery of objects. Spearfishing aqualungers are not allowed to practice their art in any underwater parks (Pennecamp in Florida, Buck Island Reef National Monument in the U.S. Virgin Islands, France's Port Cros park, and Kenya's underwater reserve) or in Mexico's Scammon Lagoon or off the Coronado Islands. Spearing is severely restricted in the Bahamas and forbidden from Bermuda and Bonaire islands. Only breath-holding divers are allowed to spearfish along the French coastlines, just in certain sections, and they are allowed only certain kinds of fish; picking lobster, shellfish, coral, ascidians, and urchins by hand or spear is prohibited. In the United States, spearfishing is not allowed in freshwater lakes in state and national parks. In a few states only trash fish such as gars and carp may be taken. All of the states bounded by an ocean allow some spearfishing, but the rules vary from place to place and time to time.

Restrictions on removal of animals other than fish also vary according to region. In the state of Maine, for instance, lobstering by divers is prohibited entirely. In Massachusetts it is a licensed activity, and off New Jersey it is unrestrained. In all cases the animals must be of legal size and cannot be egg-bearing females.

Divers planning to visit a new area should check with a game warden, the fish and game department, or a dive shop to learn local laws. Where licenses are required, often they may be obtained at the dive shop.

*Hanging suspended in the **Tongue of the Ocean,** these divers are swimming along the edge of the Bahamas, marked by a vertical cliff extending a few thousand to the deep sea bottom.*

Finders Keepers?

The recovery of objects from the water can involve a diver in considerable litigation, unless the salvage is made with the appropriate written agreements. Salvage and treasure laws vary from place to place much as spearfishing rules. Many governments have passed laws that forbid the taking of artifacts from an underwater site and impose stiff penalties upon offenders. This is particularly true where a wreck is of archaeological or historical importance. A diver taking an amphora from the bottom of the Mediterranean (which is strictly forbidden) might be removing the last clue to a ship carrying a priceless ancient cargo.

Admiralty law has a well-defined terminology for salvors. A *lost vessel* is a vessel that the owner has lost control over, against his will, either through capture, a storm, a collision, or foundering. The owner may or may not know the condition of the ship or if it still exists. A *wrecked vessel* is one that has been stranded or has run upon reef or shore with temporary, partial, or total loss. A *derelict* is a vessel or any personal property which has been discarded or abandoned by its owner in such a way as to indicate that it will not be reclaimed.

The term *salvage* can be defined in several ways, but it is generally concerned with sea-surface activity. It is the pay or reward to tugboat operators who take a disabled vessel in tow and bring her safely into port. It is awarded to persons who voluntarily save or recover a ship or cargo either in whole or in part. A ship or cargo is said to be subject to salvage when it is underwater.

According to English law, full title to wrecks, either floating or run aground, is taken by the king or the king's assignee, and title can be reclaimed by the original owner within a year and a day after the salvage has been effected. In the United States the owner of the shore against which the ship rests has title to a wreck, but the property is subject to reclamation by the original owners or the ship's insurance underwriters within a year and a day after salvaging by the shoreline owner or another salvor.

Title to sunken wrecks and cargoes remains with its original owners, his heirs, successors, or assignees. Similarly, items cast overboard during storms to lighten a distressed ship are not considered derelict. In the case of an ancient wreck whose ownership is impossible to ascertain, full title can be claimed by the state if the wreck is in waters over which it has jurisdiction. If the wreck is in international waters, its title can be claimed by anyone able to salvage it.

Derelicts, being abandoned, cannot be reclaimed by the original owner; they can be removed by the secretary of the army if they stand in the way of waterborne traffic. Salvors may work on them if the Coast Guard does not object to having a salvage boat anchored over the wreck or divers in the water at that location.

If divers are successful in recovering goods subject to import duties from a derelict ship that has been sunk in the territorial waters of the United States for more than two years, they may bring the goods through U.S. customs free of duty. Normal duty is charged for goods immersed less than two years.

With an identifying object (ship's bell, artifacts, or wood that can be dated by carbon-14 tests) he can easily raise from a wreck, a diver may be able to trace the history of his find. Before mounting a costly full-scale salvage operation, he should seek legal aid to establish ownership, and clearly detail terms for division of salvaged goods. Where local law applies, the diver should obtain a schedule for appraisals. Hard as it is, finding a treasure is easier than keeping it.

Requirements for Diving Certification

A candidate for diving certification must meet the following requirements set recently by the American Standards Institute in co-operation with representatives of each of the major U.S. organizations involved in sport diver instruction.

The minimum age for full certification is 15 years (younger students may be issued junior cards); there is no maximum age. The candidate should be able to swim 200 yards without stopping and without fins, and he should be able to stay afloat or tread water for 10 minutes without accessories. He should demonstrate his ability by completing a formal aquatic course involving both classroom and water activities.

He should be certified as medically fit for diving by a physician who understands the unique physical and emotional stresses. Diving often requires heavy exertion and the diver must be in good health, good condition, free of cardiovascular and respiratory diseases. All his body cavities must be able to readily equalize pressure. Ear or sinus pathology may impair equalization or may be aggravated by pressure. Obstructive lung disease may cause serious accidents in an ascent. A diver must not be subject to syncope, epileptic episodes, diabetic problems, or any other condition that may cause even momentary impairment of consciousness underwater. Any of these could be the cause of a diver's death. A diver must be emotionally stable. Neurotic or panicky behavior,

The sea attracts divers, but they must be careful because they have not adapted to life underwater: they remain ***visitors.***

recklessness, or proneness to having accidents may seriously endanger the diver and his companions.

The diver must understand his equipment and its maintenance. He must understand the use of face mask, fins, snorkel, inflatable flotation vests, exposure suits, weights, belt, float, flag, and knife. He must have some knowledge of tanks, demand regulators, pressure gauges, backpacks, and quick-release harnesses. He should know how to use his compass and depth gauge. The diver should also understand the physics of diving and how it applies to him. This includes buoyancy, acoustics, vision, heat loss, and gas laws, particularly as they apply to air consumption.

Candidates for diver certification should demonstrate an understanding of medical problems related to diving. These include causes, symptoms, and treatment as well as prevention of the direct effects of pressure—those that affect ears, sinuses, lungs, teeth, stomach, and intestines in the descent, and those involved in gas expansion and in lung overexpansion in the ascent. He should have similar knowledge of the indirect effects of pressure: decompression sickness, nitrogen narcosis, anoxia, and the toxicity of carbon dioxide, oxygen, and carbon monoxide. He should be familiar with other hazards such as fatigue and exhaustion, exposure, drowning, cramps, injuries from marine life, shallow-water blackout, voluntary and involuntary hyperventilation.

He should be skillful at diving entry, descent and ascent, pressure equalization, underwater swimming, mask clearing, mouthpiece clearing, buddy system, underwater and surface buoyancy control with a vest, underwater problem solving, full-gear surface snorkel swimming, weight-belt ditching, underwater gear removal and replacement, and emergency ascent.

He should be familiar with diving tables and know how to use them to solve the problems of decompression and repetitive dives. He should also have the ability to determine the limits of no-decompression dives.

He should have both specific and general knowledge of his diving environment—water conditions such as temperature, clarity, density, and movements (surface action, tides, currents); topography of bottoms and shorelines; and the marine and aquatic life, both plant and animal.

The candidate for certification should demonstrate knowledge of dive planning, use of equipment, communications both above and below water, surface survival techniques, rescue techniques, and diving-related first aid. He should be able to follow safe procedures while diving from both shore and boat and be capable of applying all his skills and knowledge to diving in the open sea.

Decompression

If a diver goes too deep, stays down too long, or comes to the surface too fast, he is subject to a number of physical disorders that could result in death. Decompression sickness, or the bends, results when the nitrogen dissolved in the diver's bloodstream and tissues under pressure comes out of solution as the diver ascends to a higher level where the ambient pressure is much less. Bubbles often lodge in limb joints, muscles, and other constricted passages, causing severe pain and paralysis. A rapid ascent can also force air out of the lungs and into the blood stream. These bubbles can lodge in the brain and cause permanent damage to the body. Dr. S. Harold Reuter has devised a set of dive charts from the complicated dive tables in the U.S. Navy's decompression manual; these direct readout charts tell clearly how deep a diver can go for how long, no-decompression limits, and decompression times.

Standard Air Decompression Table (simplified for the sport diver)

Depth (feet)	Bottom Time (min)	Decompression stops (min)		Repetitive Group
		20 (ft)	10 (ft)	
40	200		0	(*)
	210		2	N
	230		7	N
50	100		0	(*)
	110		3	L
	120		5	M
	140		10	M
	160		21	N
60	60		0	(*)
	70		2	K
	80		7	L
	100		14	M
	120		26	N
	140		39	O
70	50		0	(*)
	60		8	K
	70		14	L
	80		18	M
	90		23	N
	100		33	N
	110	2	41	O
	120	4	47	O
	130	6	52	O
80	40		0	(*)
	50		10	K
	60		17	L
	70		23	M
	80	2	31	N
	90	7	39	N
	100	11	46	O
	110	13	53	O
90	30		0	(*)
	40		7	J
	50		18	L
	60		25	M
	70	7	30	N
	80	13	40	N
	90	18	48	O
100	25		0	(*)
	30		3	I
	40		15	K
	50	2	24	L
	60	9	28	N
	70	17	39	O
110	20		0	(*)
	25		3	H
	30		7	J
	40	2	21	L
	50	8	26	M
	60	18	36	N
120	15		0	(*)
	20		2	H
	25		6	I
	30		14	J
	40	5	25	L
	50	15	31	N
130	10		0	(*)
	15		1	F
	20		4	H
	25		10	J
	30	3	18	M
	40	10	25	N
140	10		0	(*)
	15		2	G
	20		6	I
	25	2	14	J
	30	5	21	K
150	5		0	C
	10		1	E
	15		3	G
	20	2	7	H
	25	4	17	K
	30	8	24	L
160	5		0	D
	10		1	F
	15	1	4	H
	20	3	11	J
	25	7	20	K
170	5		0	D
	10		2	F
	15	2	5	H
	20	4	15	J
180	5		0	D
	10		3	F
	15	3	6	I
190	5		0	D
	10	1	3	G
	15	4	7	I

***See No Decompression Limits in "No Calculation Dive Tables" for Repetitive Groups.**

The practicality of these charts results from their being small and concise. Some are even mounted in plastic, allowing the diver to take them to the bottom as part of his gear. They could save his life.

To use the standard air decompression table on the left, a diver must know the following: all decompression stops are in minutes; ascent rate is 60 feet per minute; the chest level of the diver being decompressed should be maintained as close as possible to each decompression depth for the number of minutes stated on the chart; and the time at each stop is the exact time that is spent at that decompression depth.

For a no-decompression dive, a diver uses the table on the next page in the following way: (1) he moves along the top of the table to find the depth he has dived to, (2) drops down to the figure that denotes his bottom time, (3) goes across to the right of the table, (4) follows the arrow upward until he finds the time spent out of water since his last dive, (5) goes across to the right to find the allowable bottom time for his next dive (white numbers; they are listed under the appropriate depths at the top of each column), (6) if the no-decompression limits are exceeded, the diver must consult the decompression table for stops and time.

To use the no-calculation dive tables, one must be familiar with the following terms: *bottom time* (in minutes) is calculated as the total time from the moment that the diver leaves the surface until the moment the diver starts a direct ascent back to the surface; *depth* (in feet) is the deepest point of descent; *residual nitrogen time* (in minutes) is the time that a diver should consider he has already spent on the bottom when he starts a repetitive dive; *repetitive dive* is a dive begun within 12 hours of surfacing from a previous dive.

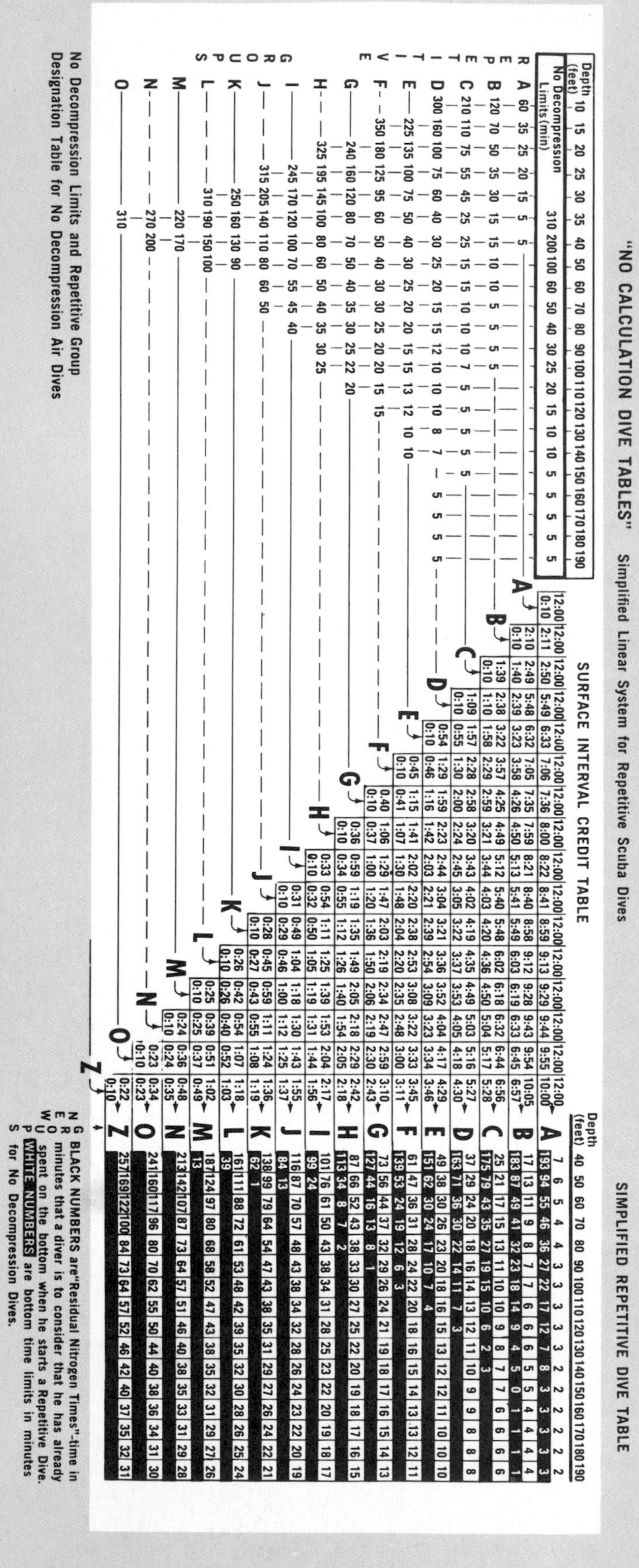

Repetitive Dives

The no-calculation dive tables are also used to indicate how long and how deep divers can go on a second, third, or fourth repetitive dive without having to go through a lengthy process of decompression upon returning to the surface. For example, suppose a diver descends 50 feet and remains there 60 minutes (bottom time). He then returns to the surface for 3 hours (surface interval) before beginning his next dive. He would like to know how long he can stay on the bottom at 70 feet during his second dive (first repetitive dive) without having to worry about decompressing when he surfaces. He uses the table in the following manner to find the answer: (1) he locates the 50-foot column on the left side of the table and drops down to the 60-minute level (H group), (2) goes across to the right to the H group and follows the arrow upward to locate the time interval in which 3 hours falls (2:24 to 3:20), (3) goes across to the right to the D group, to the 70-foot column, where the bottom time limit is indicated at 30 minutes. This is the maximum time he can spend at 70 feet during his second dive (first repetitive dive) without having to undergo decompression.

Suppose the diver spent only 15 of his 30 minutes of bottom time at 70 feet and then returned to the surface, where he remained for 1 hour before beginning his second repetitive dive, this time to a depth of 50 feet. To calculate how long he can stay at 50 feet without having to decompress upon surfacing: (1) the diver would add the 15 minutes of bottom time used during his first repetitive dive at 70 feet to the 20 minutes of residual nitrogen time—this is a total of 35 minutes. (2) The diver now returns to the left-hand table, drops down the 70-foot column to 35 minutes, and follows the procedure outlined above.

"NO CALCULATION DIVE TABLES" Simplified Linear System for Repetitive Scuba Dives

Depth (feet)	10	15	20	25	30	35	40	50	60	70	80	90	100	110	120	130	140	150	160	170	180	190
No Decompression Limits (min)						310	200	100	60	50	40	30	25	20	15	10	10	5	5	5	5	5
A	60	35	25	20	15	5	5															
B	120	70	50	35	30	15	15	10	10	5	5	5	5									
C	210	110	75	55	45	25	25	15	15	10	10	10	7	5	5	5	5	5				
D	300	160	100	75	60	40	30	25	20	15	15	12	10	10	10	8	7	—	5	5	5	5
E		225	135	100	75	50	40	30	25	20	20	15	15	13	12	10	10					
F		350	180	125	95	60	50	40	30	30	25	20	20	15	15							
G			240	160	120	80	70	50	40	35	30	25	22	20								
H			325	195	145	100	80	60	50	40	35	30	25									
I				245	170	120	100	70	55	45	40											
J				315	205	140	110	80	60	50												
K					250	160	130	90														
L					310	190	150	100														
M						220	170															
N						270	200															
O						310																

REPETITIVE GROUPS

No Decompression Limits and Repetitive Group Designation Table for No Decompression Air Dives

SURFACE INTERVAL CREDIT TABLE

	A	B	C	D	E	F	G	H	I	J	K	L	M	N	O	Z
A	12:00 0:10	12:00 2:11	12:00 2:50	12:00 5:49	12:00 6:33	12:00 7:06	12:00 7:36	12:00 8:00	12:00 8:22	12:00 8:41	12:00 8:59	12:00 9:13	12:00 9:29	12:00 9:44	12:00 9:55	12:00 10:00
B		2:10 0:10	2:49 1:40	5:48 2:39	6:32 3:23	7:05 3:58	7:35 4:26	7:59 4:50	8:21 5:13	8:40 5:41	8:58 5:49	9:12 6:03	9:28 6:19	9:43 6:33	9:54 6:45	10:05 6:57
C			1:39 0:10	2:38 1:10	3:22 1:58	3:57 2:29	4:25 2:59	4:49 3:21	5:12 3:44	5:40 4:03	5:48 4:20	6:02 4:36	6:18 4:50	6:32 5:04	6:44 5:17	6:56 5:28
D				1:09 0:10	1:57 0:55	2:28 1:30	2:58 2:00	3:20 2:24	3:43 2:45	4:02 3:05	4:19 3:22	4:35 3:37	4:49 3:53	5:03 4:05	5:16 4:18	5:27 4:30
E					0:54 0:10	1:29 0:46	1:59 1:16	2:23 1:42	2:44 2:03	3:04 2:21	3:21 2:39	3:36 2:54	3:52 3:09	4:04 3:23	4:17 3:34	4:29 3:46
F						0:45 0:10	1:15 0:41	1:41 1:07	2:02 1:30	2:20 1:48	2:38 2:04	2:53 2:20	3:08 2:35	3:22 2:48	3:33 3:00	3:45 3:11
G							0.40 0:10	1:06 0:37	1:29 1:00	1:47 1:20	2:03 1:36	2:19 1:50	2:34 2:06	2:47 2:19	2:59 2:30	3:10 2:43
H								0:36 0:10	0:59 0:34	1:19 0:55	1:35 1:12	1:49 1:26	2:05 1:40	2:18 1:54	2:29 2:05	2:42 2:18
I									0:33 0:10	0:54 0:32	1:11 0:50	1:25 1:05	1:39 1:19	1:53 1:31	2:04 1:44	2:17 1:56
J										0:31 0:10	0:49 0:29	1:04 0:46	1:18 1:00	1:30 1:12	1:43 1:25	1:55 1:37
K											0:28 0:10	0:45 0:27	0:59 0:43	1:11 0:55	1:24 1:08	1:36 1:19
L												0:26 0:10	0:42 0:26	0:54 0:40	1:07 0:52	1:18 1:03
M													0:25 0:10	0:39 0:25	0:51 0:37	1:02 0:49
N														0:24 0:10	0:36 0:24	0:48 0:35
O															0:23 0:10	0:34 0:23
Z																0:22 0:10

SIMPLIFIED REPETITIVE DIVE TABLE

Depth (feet)	40	50	60	70	80	90	100	110	120	130	140	150	160	170	180	190
A	7	6	5	4	4	3	3	3	3	3	2	2	2	2	2	2
	193	94	55	46	36	27	22	17	12	7	8	3	3	3	3	3
B	17	13	11	9	8	7	7	6	6	6	5	5	4	4	4	4
	183	87	49	41	32	23	18	14	9	4	5	0	1	1	1	1
C	25	21	17	15	13	11	10	10	9	8	7	7	6	6	6	6
	175	79	43	35	27	19	15	10	6	2	3					
D	37	29	24	20	18	16	14	13	12	11	10	9	9	8	8	8
	163	71	36	30	22	14	11	7	3							
E	49	38	30	26	23	20	18	16	15	13	12	12	11	10	10	10
	151	62	30	24	17	10	7	4								
F	61	47	36	31	28	24	22	20	18	16	15	14	13	13	12	11
	139	53	24	19	12	6	3									
G	73	56	44	37	32	29	26	24	21	19	18	17	16	15	14	13
	127	44	16	13	8	1										
H	87	66	52	43	38	33	30	27	25	22	20	19	18	17	16	15
	113	34	8	7	2											
I	101	76	61	50	43	38	34	31	28	25	23	22	20	19	18	17
	99	24														
J	116	87	70	57	48	43	38	34	32	28	26	24	23	22	20	19
	84	13														
K	138	99	79	64	54	47	43	38	35	31	29	27	26	24	22	21
	62	1														
L	161	111	88	72	61	53	48	42	39	35	32	30	28	26	25	24
	39															
M	187	124	97	80	68	58	52	47	43	38	35	32	31	29	27	26
	13															
N	213	142	107	87	73	64	57	51	46	40	38	35	33	31	29	28
O	241	160	117	96	80	70	62	55	50	44	40	38	36	34	31	30
Z	257	169	122	100	84	73	64	57	52	46	42	40	37	35	32	31

NEW GROUPS

BLACK NUMBERS are "Residual Nitrogen Times"-time in minutes that a diver is to consider that he has already spent on the bottom when he starts a Repetitive Dive. WHITE NUMBERS are bottom time limits in minutes for No Decompression Dives.

Recompression

Sometimes for one reason or another a diver is forced to make a rapid ascent, disregarding the decompression stops he should make. Such a situation could arise if the diver's breathing system malfunctioned, if his air supply ran dangerously low so as not to leave him enough time to gradually surface, or if he was injured and needed immediate medical attention. But, for whatever the reasons, an uncontrolled ascent will result in decompression sickness and must be treated immediately. If a recompression chamber is not available at the surface, the stricken diver should lie down with his feet raised. Under no circumstances should he be flown in an unpressurized airplane above 1000 feet. Such a decrease in pressure could kill him. When a chamber is at hand, the stricken diver should be placed inside and the unit should be pressurized, either mechanically on deck or by lowering the chamber to the proper depth. The chart below

Treatment of Decompression Sickness and Air Embolism

Stops		Bends—Pain only				Serious Symptoms	
Rate of descent —25 ft. per min. Rate of ascent —1 min. between stops.		Pain relieved at depths less than 66 ft. Use Table 1-A if O_2 is not available.		Pain relieved at depths greater than 66 ft. Use Table 2-A if O_2 is not available. If pain does not improve within 30 min. at 165 ft. the case is probably not bends. Decompress on Table 2 or 2-A.		Serious symptoms include any one of the following: 1. Unconsciousness. 2. Convulsions. 3. Weakness or inability to use arms or legs. 4. Air embolism. 5. Any visual disturbances. 6. Dizziness. 7. Loss of speech or hearing. 8. Severe shortness of breath or chokes. 9. Bends occurring while still under pressure.	
						Symptoms relieved within 30 min. at 165 ft. Use Table 3	Symptoms not relieved within 30 min. at 165 ft. Use Table 4
Pounds	**Feet**	**Table 1**	**Table 1-A**	**Table 2**	**Table 2-A**	**Table 3**	**Table 4**
73.4	165	....	....	30 (air)	30 (air)	30 (air)	30 to 120 (air)
62.3	140	....	....	12 (air)	12 (air)	12 (air)	30 (air)
53.4	120	....	....	12 (air)	12 (air)	12 (air)	30 (air)
44.5	100	30 (air)	30 (air)	12 (air)	12 (air)	12 (air)	30 (air)
35.6	80	12 (air)	12 (air)	12 (air)	12 (air)	12 (air)	30 (air)
26.7	60	30 (O_2)	30 (air)	30 (O_2)	30 (air)	30 (O_2) or (air)	6 hrs. (air)
22.3	50	30 (O_2)	30 (air)	30 (O_2)	30 (air)	30 (O_2) or (air)	6 hrs. (air)
17.8	40	30 (O_2)	30 (air)	30 (O_2)	30 (air)	30 (O_2) or (air)	6 hrs. (air)
13.4	30	↓	60 (air)	60 (O_2)	2 hrs. (air)	12 hrs. (air)	First 11 hrs. (air) Then 1 hr. (O_2) or (air)
8.9	20	5 (O_2)	60 (air)	↓	2 hrs. (air)	2 hrs. (air)	First 1 hr. (air) Then 1 hr. (O_2) or (air)
4.5	10	↓	2 hrs. (air)	5 (O_2)	4 hrs. (air)	2 hrs. (air)	First 1 hr. (air) Then 1 hr. (O_2) or (air)
Surface		↓	1 min. (air)	↓	1 min. (air)	1 min. (air)	1 min. (O_2)

Time at all stops in minutes unless otherwise indicated.

A fishy survey. Using a plexiglas sphere as a temporary tank, a diver in the Red Sea makes a survey. The fish will all be later released.

describes the procedures that must be strictly adhered to in treating decompression and air embolism cases. It is important to recognize all symptoms and choose the proper schedule of recompression as indicated in Tables 1 through 4. As an example, suppose a stricken diver shows signs of the bends and his symptoms are relieved at pressures equal to depths less than 66 feet. If oxygen were not available, only air, he would be decompressed according to Table 1-A. The chamber would be lowered at a rate of 25 feet per minute to a simulated depth of 100 feet, where it would remain for 30 minutes, the diver inside breathing air. After the prescribed 30 minutes, the chamber would be raised to 80 feet (1 minute must be taken between stops), where it would stay for 12 minutes. The rest of the schedule would be followed most stringently until the diver arrived at the surface, at which time he should be cured. If symptoms reoccur during treatment, the diver would again be lowered to a depth giving him relief. If this is less than 30 feet, the chamber would be lowered to 30 feet and decompression would proceed according to Table 3. If relief from pain occurred deeper than 30 feet, the patient would be kept at that depth for 30 minutes and his decompression ultimately completed according to Table 3. If, however, his original treatment followed Table 3, then Table 4 would be used. It is safe to say that these procedures as defined have saved the lives of many divers afflicted with assorted decompression ailments. It is strongly recommended however, that precautions be taken so that recompression is not necessary.

Chapter IV. Advice from Cousteau

Within as short a time period as 30 years, man has gained free access to the sea, yet within that same period he has begun to watch the sea die. Many of the reefs that were abundant with life only 15 years ago are now almost deserts. A startling number of beautiful reefs in the warm Caribbean waters around Florida have been virtually destroyed by man's pollution. Those who have witnessed this destruction of the ocean feel compelled to recommend strong measures to be taken by national governments and private enterprise in order to alleviate the burdens of pollution and overfishing. Almost as important are the responsibilities that each one of us must live up to. The sea can only be rescued if people understand and love it. Each one of us who comes in contact with the sea, whether on shore, on the surface, or in a dive, should observe certain "do's and don'ts" in order to make the experience more complete, more enjoyable, and safer for ourselves as well as future generations.

Education is the key to our encounter with the sea. Knowledge of marine plants and animals adds to our experience underwater and enables us to gain respect for the delicate balances of life in the sea. A familiarity with the dangers in the sea, including poisonous and venomous animals, helps to eliminate the beginning diver's panic in the presence of the unknown. Never attempt to dive without first taking certified instruction and reading a good diving manual.

Recreational fishing is not a hero's game. When a big-game fisherman—with the assistance of a 1000-horsepower motorboat, the latest sophisticated rod, reel, and line, and the guidance of a local expert—finally gets photographed standing proudly near his larger-than-man catch, I am only able to compare this so-called hero—most of the time a physiological ruin—with the magnificent and hopeless victim to know where my heart is. A fish struggling for life on the deck of a boat may not appear to be suffering. Just because sea animals are mute does not mean that they do not agonize in slow death.

Many are interested in the ocean but don't dive, confining their contact with the sea to the shore. Yet when their interest in the ocean causes them to buy shells and coral jewelry, they are also contributing to the destruction of the sea, even though they did not go down in person to hack away at a reef. Do not encourage the practice of commercial diving for shells by purchasing bleached coral and shells. Even those specimens that appear in museum gift shops have usually been taken from a reef or with a living animal inside. Shells picked up on the beach are not useful to these shops since the action of the waves has often destroyed their construction or their color.

Practices that are beneficial to the ocean's survival will not interfere with the enjoyment of the sea. On the contrary, they will often bring a greater excitement to the recreational use of the ocean. For example, rare artifacts from sunken Greek and Roman galleys off the islands in the Marseilles region, went unobserved for 10 years by spear-gun-carrying divers who frequented these waters. Their concentration on pursuing and killing fish prevented them from having a once-in-a-lifetime experience of discovery. When they realized this fact, and perhaps considered how much money they may have lost, most of them threw their spears away forever.

We do not know yet how to breed artificially the tropical fish today obtained by ***raping of reefs.***

Tide Pool Tragedy

For the nondiver, the coastline offers a varied and lively display of marine animals. Ten years ago, because of new public interest in the sea, visits to tide pools became the favorite field trip for biology and elementary science classes. Well-meaning teachers would arrange an outing to coincide with low tides, and take their classes on collecting trips. Most tide pool organisms are defenseless against us. Thousands of enthusiastic school children devastated these important marine habitats, especially in southern California where the practice was common. When the catastrophic impact of those trips was recognized, the practice of unleashing youngsters on the tidal areas was stopped. This does not mean that observation of tidal areas should be forbidden; but groups have to be kept small and under control. Observation must be made from above the tidemark. Each step taken across the low tide mark on a rocky coast will crush many organisms.

Intertidal zone organisms are fascinating, but have been endangered by **too much collecting.**

Trips to tide pools were often organized as hunting excursions for the purpose of collecting animals for saltwater aquaria. Unfortunately, in the present state of technology, it is impossible to approve of these aquaria. Even the stock marine animals that are purchased from supply houses are specimens captured all around the world to meet the growing demand. Until ocean organisms can be hatched in captivity using the techniques of mariculture, fish enthusiasts should limit themselves to traditional, freshwater aquaria, the stock for which is cultured rather than taken from natural populations.

The only objects that should be picked up from a coastal area are nonliving ones that wash up with the tide. Empty shells can be collected, but living animals should not be taken from the beach except clams or other organisms whose populations are large and are regulated by fishing rules.

Sailor's Rules

Millions of pleasure boats cruise the coastal waters and open seas of the world. There are certain rules of the sea that recreational sailors should observe.

Motorboats have much greater ability to harm the sea than sailing vessels. They consume more fuel, and their noisy motors are irritating. But if owners of pleasure boats insist on outboard motors, they should purchase four-cycle rather than two-cycle motors. Two-cycle outboards burn oil in their cylinders, as well as gasoline, leaving an oily film in their wake.

Sewage should not be dumped from a boat into the sea. In shallower harbor areas, where water flow may already be impeded,

dumping is especially dangerous. Raw sewage could settle to the bottom and remain there. Its subsequent biological breakdown by bacteria and other organisms would deplete oxygen levels near the bottom and wipe out any healthy benthic community that might still exist in the area. The sewage could also become a breeding ground for viruses and bacteria potentially dangerous to man. During summer months it is not uncommon to see bathing beaches closed for that very reason. Accidentally swallowing some contaminated water could cause one severe discomfort and lead to more serious problems. At most ports there are sewage disposal systems, making dumping at sea unnecessary. It is especially cruel to drop the rings from aluminum cans or steel bottle caps over the side. As they sink, they spin, reflecting light. Their glitter attracts large fish who swallow them just as they would a hook. Often the metal perforates the stomach of the fish, killing it. It is lost; it cannot even be used as a source of food.

One of the most popular uses of pleasure boats is for inshore and offshore recreational fishing. The reason why this apparently innocent practice is becoming harmful is directly related to a much larger population, with the attendant increase in affluence and leisure time. The taking of fish for sport alone has become deplorable in most cases. Reef, rock, and coastal fish taken by many amateur fishermen has resulted in depletion of the numbers of these fish. Fishing in the open sea to satisfy simple food needs is healthy in principle. But, considering the very large number of sports fishermen in the world (9.5 million in the U.S.), even limited fishing can have an enormous effect on the ocean's fish populations.

I hope that parents will emphasize to their children that there is no relationship between recreational fishing and proof of heroism or any other romantic ideal. The idea that a man listening to his portable radio and eating sandwiches and drinking beer can prove his strength, character, and courage by engaging in a long and painful battle to the death with a hooked animal is absurd.

***Pleasure boats** lose their appeal when they flush their bilges or holding tanks into the ocean.*

Diving Do's and Don'ts

Diving is the ultimate way to enjoy the sea. A trained diver who observes certain rules has an exciting life of adventure awaiting him. My personal advice to divers should not in any way serve as a substitute for diving lessons and actual experience.

It is wise to use caution at all times in the sea, but diving with open-circuit air lungs is healthy, even for young children and the aged. My own children and grandchild started at the age of four. My father started at 67; he dived until he was 82 and died at 92. During a dive at 130 feet, the partial pressure of oxygen in inhaled air is the same as if pure oxygen were breathed at the surface; actually this is good for the heart muscles, temporarily at least. Naked diving, however, imposes an exhausting strain on the circulatory and respiratory systems; it should only be undertaken by young adults in peak athletic condition.

The only inconveniences, even in shallow dives, are to the sinuses and ears. The pain is a warning that your pressure-compensating systems are not working well. Don't insist—don't panic—don't go back to the surface. Just stay there at the border of pain; relax, blow your nose into your mask, swallow several times. After three or four careful dives, you will forget your ears. Cold water is a physiological strain on the diver. Protect yourself from this stress by wearing an insulated suit. Body heat must be replaced as soon as you surface. It is too much to ask of your body to warm itself from the inside out. Take a warm bath, or light a beach fire, and have a cup of hot tea.

Sharks have the reputation of being the worst enemies of the diver. Many species of shark potentially deserve that reputation—they are fast swimmers and have in their mouth the most effective weapon in the sea. But usually they present a small threat to the experienced diver. Sharks are generally cautious; the diver has ample time to assess the danger and to make a decision. Be especially careful when entering and leaving the water, because the only place where sharks attack without hesitation is at the surface. Water skiing in shark-infested areas, for example, is more dangerous than diving in these same locales. A single shark poses no serious threat to the diver. Keep him in view and be prepared to push your camera forward if he wanders too close. A pack of sharks is a different situation. No matter what you do, one will always be behind you, out of sight. Get out of the water *without* haste, remembering again that you are most vulnerable as you cross the surface. If you meet a large, light-gray shark with five gill slits and a homocercal (symmetrical) tail, it is also best to get out of the water, while observing the aforementioned precautions: the animal may be the great white shark *(Carcharodon carcharias),* the most dreaded of them all. But less dramatically, sea urchins, fire coral, and stinging jellyfish are much more of a daily nuisance than sharks. Poisonous animals one must be cautious about are discussed at length in another chapter in this volume.

I warn divers about the inflatable life-jacket. Dangerous for beginners, its use often heightens rather than prevents danger. An inexperienced diver is often prone to fear. A new environment, filled with foreign animals, amazes but also breeds caution. An unfamiliar animal or any diving irregularity makes the beginner want to rush back to the surface. The inflatable vest will balloon him up very quickly. If he is frightened or tense, his breathing mechanism may tighten. Air in the lungs will expand as pressure decreases, and air embolism may result. Nine out of ten diving deaths occur at the surface. Learn about your equipment but don't overburden yourself with gadgets; I have seen divers that look like aquatic Christmas trees.

Enter the sea gently, don't jump into the water. The critical layer is the first three to ten feet beneath the surface, where the relative varieties of pressure are the greatest and the most decisive for the rest of the dive. Use a ladder to enter the water very slowly. If you intend to make more than one dive in a day, make the first one the deeper, the second one more shallow. If you started with, for example, a long shallow dive in only 35 feet without the need of decompression, you still would have dissolved an appreciable quantity of nitrogen in your tissues; a following deep dive would no longer be safe according to the conventional decompression tables. This fact is responsible for many very severe decompression accidents.

Take the time to adjust your belt weights in order to be neutrally buoyant. Then stay in midwater; never step on the bottom, as it may be paved with stinging creatures.

Kelp beds are an intriguing habitat to explore. Yet, there can be danger for the careless diver; it is easy to get entangled in strands of algae that may be as long as 100 feet. When diving in a kelp bed don't allow yourself to run too low on air, which would require an ascent through the tangled canopy. Also make sure when ascending near the bed that you look upward to avoid any stray kelp fronds. Be sure to take along a sharp knife; you may need it to cut yourself free. If you get tangled, don't panic and don't thrash about. Like a fly in a spider's web, you'll only tangle yourself more. Don't slash blindly with your knife. Use it slowly, cutting away the fronds around your regulator, weight belt, or whatever. Patience in such a situation could mean the difference between life and death for you.

I mentioned in Volume XVII, *Riches of the Sea,* that I am opposed to the use of spear guns. This is not only because I shudder at the loss the sea suffers from the guns presently in use but also because carrying a spear and hunting can detract from the real pleasures and values of diving. Abandoning the gun will free your attention to observe all that is around you and will free your hand to carry a camera. Photographs of a diving experience are more than lasting souvenirs of adventure. They facilitate identification of animals and will contribute much more to your knowledge of the ocean than a dead specimen disfigured beyond recognition on the end of a diver's spear.

__Diving__ is not a frivolous sport; one must be aware of the hazards involved and act accordingly.

Chapter V. Poisonous and Venomous Marine Animals*

The study of poisonous and venomous marine animals is called *marine zootoxicology.* At present very little is known about the subject of poisonous marine plants. The term *poisonous* is restricted here to those organisms that are poisonous to eat. A *venomous* animal is one that can produce a sting. The venom apparatus of these animals consists of a wound-producing organ (a tooth or spine, for example) and a specialized venom gland which produces the poison. The wound-producing organ penetrates the flesh of the victim and delivers the poison.

Poisonous Marine Animals

The number of poisonous marine animals is vast, and new toxic species are frequently discovered. Toxic marine organisms are most prevalent in tropical and warm temperate waters, but representative species are found in every sea of the world. They range from the one-celled protozoa, through invertebrates and fishes, to giant marine mammals in the form of whales and polar bears.

Dinoflagellates

Because of their plant-animal characteristics, dinoflagellates are classed as plants by botanists and animals by zoologists. Regardless of their classification, they are microscopic one-celled creatures. Some dinoflagellates cause the "red tide," a phenomenon that occurs from time to time in bays or other coastal waters, usually in warm weather. These organisms multiply so rapidly and to such an extent that the water becomes discolored by their presence. This condition may be fatal to large numbers of the local fish population, because these tiny organisms produce potent poisons.

**Condensed from the works of Bruce W. Halstead.*

Man can be poisoned by dinoflagellates when he eats mussels, clams, oysters, or other shellfish that have fed on the poisonous organisms. The resulting disease is *paralytic shellfish poisoning;* its symptoms include gastrointestinal disturbances, numbness, paralysis, salivation, and difficulty in swallowing. There is no specific treatment or antidote presently available.

The extreme toxicity of this poison cannot be overemphasized. Usually there are quarantine regulations governing the sale or eating of toxic shellfish in endemic areas such as the west coast of the United States and Canada, northeast Canada, Norway, England, Germany, Belgium, the Netherlands, France, South Africa, and New Zealand.

Fish

Man can experience four types of fish poisoning: ciguatera and scombroid, puffer, and clupeoid poisoning. Toxic fish, which are most frequently encountered in tropical or warm temperate waters, can have a serious, perhaps even fatal, effect on man.

Ciguatera. At least 300 different species of fish can cause ciguatera. It is believed that these fish become poisonous because of what they eat. It seems that under the proper circumstances almost any species of marine fish may cause this type of poisoning. The poisonous flesh of the fish does not result from bacterial contamination but is due to a nerve poison known as ciguatoxin. Ciguatera is often a serious health problem in the central and south Pacific Ocean and the West Indies. It also occurs in parts of the Indian Ocean.

***Scorpionfish** are a group of highly venomous marine animals found in tropical, temperate, and arctic waters. Their sting causes severe pain and in some cases can be fatal if not treated immediately.*

Ciguatoxic fish include: jack (tropical Atlantic); surmullett (Indo-Pacific); seabass (West Indies); surgeonfish (Indo-Pacific); snapper (Indo-Pacific); red snapper (Indo-Pacific); grouper (Indo-Pacific); sea bass (Indo-Pacific); barracuda (circumtropical); and moray eel (Indo-Pacific).

Symptoms of ciguatera include tingling sensations and numbness about the lips, tongue, throat, and extremities, as well as skin rashes and itching. Gastrointestinal upset may be present. There is frequently a feeling of extreme weakness. A characteristic symptom is the "dry ice or electric shock" sensation—hot objects may feel cold and cold objects feel hot. In severe cases there may be muscular incoordination, paralysis, and convulsions, and death may result.

Treatment and prevention. There is no specific treatment or antidote for ciguatera. The stomach should be emptied as soon as possible. Warm salt water or egg white is effective as an emetic. Placing a finger down the throat will induce vomiting. Since the treatment is largely symptomatic, the victim should be hospitalized as soon as possible.

The appearance of a fish gives no indication of its toxicity. But a few general rules should be followed to avoid ciguatera. A very large specimen is more likely to be toxic than a small specimen of the same species. The viscera—liver, intestines, roe—of tropical fish should never be eaten: these are the most dangerous portions of the fish. Barracuda, jacks, and groupers should not be eaten during their reproductive seasons. The toxic properties of the fish are not affected by methods of either preservation or cooking. Local advice should be sought as to safety of eating a particular type of tropical fish.

Scombroid poisoning. Under most circumstances most scombroid fish—tuna, bonito, mackerel, skipjack, among others—are edible. However, toxicity can develop when the fish is not properly preserved. Fish normally contain a chemical constituent in their flesh called histidine. When histidine is acted upon by certain types of bacteria, it changes into a histaminelike substance known as saurine. This substance is produced when scombroid fish stand at room temperature or in the sun for several hours.

The symptoms of scombroid poisoning resemble those of a severe allergy—intense headache, dizziness, throbbing of the large vessels of the neck, dryness of the mouth, thirst, palpitation of the heart, difficulty in swallowing, nausea, vomiting, diarrhea, and intense itching. Death may occur, but is rare. Generally the acute symptoms last only about 10 hours, and a rapid recovery follows.

Treatment and prevention. Treatment of scombroid poisoning is symptomatic. The victim's stomach should be evacuated. Any of the ordinary antihistaminic drugs will relieve the victim's discomfort. Prevention of this type of poisoning is simple—fish left unrefrigerated for longer than two hours should be discarded.

Puffer poisoning. The group of fish known as puffers can take in large quantities of air or water, thus swelling themselves up like balloons. There are over 90 different species of puffers. These fish are among the most poisonous of all marine creatures. Their liver, gonads, intestines, and skin usually contain a powerful nerve poison known as tetrodotoxin. The origin of tetrodotoxin is not known, but its presence is influenced to some extent by the reproductive cycle of the fish.

Although puffers are most numerous in the tropics, many are also found in temperate zones. Puffers are easily recognized by their characteristic boxy shape and four large platelike teeth.

Symptoms of puffer poisoning include tingling about the lips and tongue and loss of

motor coordination. The entire body may become numb, and the victim may feel that he is "floating." Excessive salivation, extreme weakness, nausea, vomiting, diarrhea, abdominal pain, twitching of the muscles, paralysis, difficulty in swallowing, loss of voice, convulsions, and death by respiratory paralysis may ensue. More than 60 percent of the victims poisoned by this fish die. The poison is not destroyed by cooking.

Treatment and prevention. There is no specific antidote for puffer poisoning. The treatment is similar to that of ciguatera.

Despite its great toxicity, this fish commands the highest prices in Japan as a food fish. Puffers, which are known as *fugu* in Japan, are prepared and sold in special restaurants, which hire specially trained *fugu* cooks who are licensed to handle this deadly fish. The *fugu* must be carefully prepared in a very special way or it may cause violent death. But the wisest course for all but the *fugu* cook is to learn to recognize the puffer fish and leave it alone.

Clupeoid poisoning. This form of poisoning is caused by the ingestion of certain types of herringlike fish in certain parts of the tropical Pacific, West Indies, and Indian Ocean. The origin of the poison is unknown, but is believed to come from the food web of the fish. Symptoms of clupeoid poisoning include nausea, vomiting, abdominal pain, coma, and convulsions. Death may occur within less than 30 minutes. It is an extremely violent and rapid-acting poison. Fortunately it is a rare form of poisoning, since the fish are toxic only sporadically. The poison is not destroyed by cooking.

Treatment and prevention. There is no specific antidote for clupeoid poisoning, but

Respiratory paralysis, induced by ***pufferfish tetrodotoxin,*** *kills about 20 people in Japan each year.*

treatment is similar to that for ciguatera. Visitors to the areas where this poisoning occurs should check with local fishermen concerning the edibility of tropical herrings.

Marine Turtles

Poisoning from marine turtles is one of the least known forms of intoxications produced by marine organisms. Certain species of marine turtles—the green turtle, hawksbill turtle, and leatherback turtle in the Philippines, Ceylon, India, and Indonesia—may be extremely toxic. Their poison appears to result from something in their diet which does not harm them but makes the flesh toxic. The poison is not destroyed by heat or cooking.

Symptoms of turtle poisoning include gastrointestinal upset, dizziness, and a dry burning sensation of the lips, tongue, and lining of the mouth and throat. Swallowing may become very difficult, and excessive salivation may occur. The tongue may develop a white coating, and the breath may become very foul. The tongue may later become ulcerated. In severe cases the victim becomes sleepy and gradually loses consciousness. Death may result. Almost half of the people poisoned by marine turtles die. There is no specific treatment or antidote.

In the tropical Indo-Pacific region marine turtles should be eaten with extreme caution. Visitors to that area should check first with the local public health authorities. Turtle liver is especially dangerous to eat.

Polar Bear

Numerous people have been poisoned as a result of having dined on the liver and kidneys of polar bears, but fatalities have been rare indeed. And since the polar bear is considered to be an endangered species, we hope that no one will join a hunting party and kill or add polar bear to their menu.

Venomous Marine Animals

Hydroids, Jellyfish, Corals, and Anemones

Hydroids are a group of coelenterates that have well-developed polyps. They may be solitary or colonial, and they usually reproduce by budding, thus forming free-swimming medusae. There are about 2700 hydroid species. Because of their sting, the fire coral (*Millipora* spp.) and the Portuguese man-of-war (*Physalia* spp.), which are found in most tropical seas, are dangerous to divers.

Jellyfish are the group (Scyphozoa) that includes the true medusae, which are characterized by eight notches in their bell. The deadly sea wasps (*Chironex fleckeri; Chiropsalmus* spp.) and the jellyfish (*Cyanea* spp.) are members of this group. There are about 200 species of jellyfish found throughout the seas of the world.

An example of stinging coral is the Elk Horn coral (*Acropora palmata*). Although many corals are capable of inflicting nasty wounds, most of them do not produce venom. There are a number of sea anemones, such as *Actinia*, *Actinodendron*, and *Anemonia*, that are capable of inflicting venomous wounds.

Dinoflagellates, *in great numbers, produce a toxin which makes some filter feeders poisonous.*

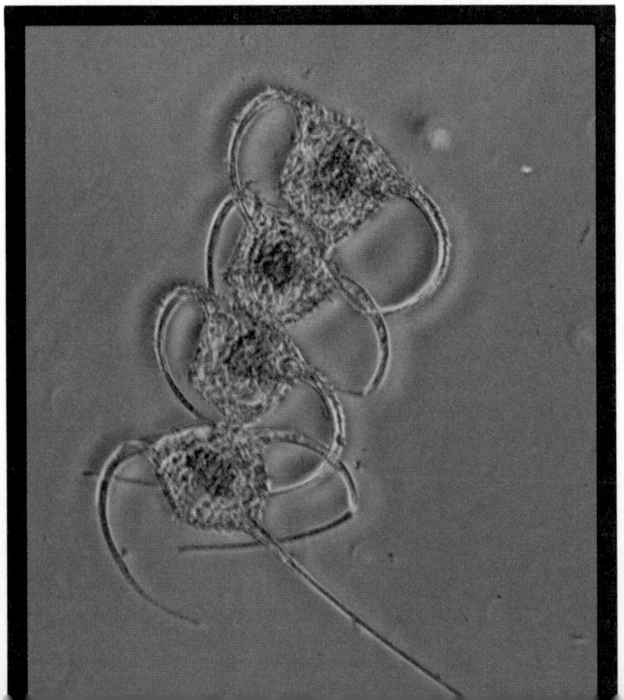

The venom apparatus of coelenterates consists of the nematocysts, or stinging cells, which are largely located within the outer layer of tissue of their tentacles. When a diver comes in contact with the tentacles of a coelenterate, he brushes up against the cnidocils of literally thousands of these minute stinging structures, and thousands of small doses of venom are injected.

The symptoms produced by coelenterate stings are primarily local skin irritations. However, in the case of *Physalia,* the sting may be extremely painful and in rare cases may cause death. While the stings of many jellyfish may be only annoying, stings from *Cyanea* and *Chiropsalmus* may produce severe reactions. *Chironex* stings are extremely dangerous and can cause almost instant death. Fortunately, the geographical range of *Chironex* appears to be limited to the northeast Australian coast.

A coelenterate sting may cause an immediate mild prickly or stinging sensation like that of a nettle sting. Or it may produce a burning, throbbing, or shooting pain which may render the victim unconscious. The area that came in contact with the tentacles usually becomes reddened; a severe rash, blistering, swelling, and minute skin hemorrhages may follow. In severe cases, in addition to shock there may be muscular cramps, abdominal rigidity, diminished touch and temperature sensation, nausea, vomiting, backache, loss of speech, frothing at the mouth, sensation of constriction of the throat, respiratory difficulty, paralysis, delirium, and convulsions. Death may follow.

Mussels, *if contaminated by toxic dinoflagellates, cause* ***paralytic shellfish poisoning*** *in humans.*

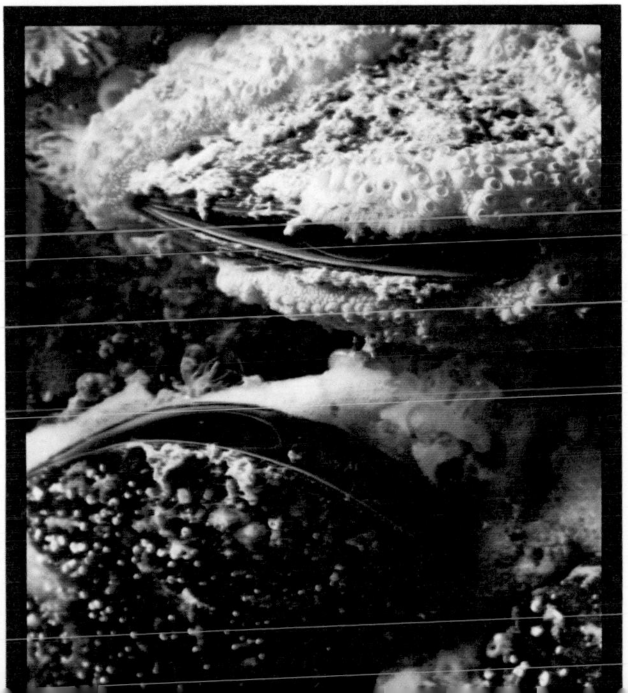

Upon being stung, the victim should make every effort to get out of the water as soon as possible because of the danger of drowning. The tentacles of some jellyfish may cling to the skin. Care should be exercised in the removal of the tentacles or additional stings will be received. A towel, rag, seaweed, stick, or handful of sand should be used. Dilute ammonia and alcohol should be applied to the site of the sting as soon as possible.

Treatment and prevention. Oral histaminics and topical cream are useful in treating the rash caused by the sting. Artificial respiration, cardiac and respiratory stimulants, and other forms of supportive measures may be needed, thus requiring the services of a physician who may have to use morphine and intravenous injections of calcium gluconate. There are no specific antivenins available.

Jellyfish tentacles may trail 50 feet or more behind the organism. The diver should therefore give jellyfish a wide berth. Tight-fitting woolen or nylon underwear and rubber diving suits are useful in affording protection from attacks from these creatures. Jellyfish washed up on the beach, though appearing dead, may be capable of inflicting a serious sting. Swimming immediately after a tropical storm may result in multiple stings from remnants of damaged tentacles floating in the water.

Coral cuts can be a serious problem in the tropics. The primary reaction to a coral cut, sometimes called "coral poisoning," is the appearance of red welts and itching around the wound. If coral cuts are left untreated, a mere scratch may become a painful ulcer. Coral cuts should be promptly cleansed and painted with an antiseptic solution. In severe cases, a physician is needed, and it may be necessary to give the victim bed rest with elevation of the limb. Kaolin poultices, dressings of magnesium sulfate in glycerin solution, and antibiotics may also be needed. Antihistaminic drugs are also useful.

Annelid Worms

Annelid worms are elongate organisms whose body is usually segmented, each segment bearing a pair of setae, or bristlelike structures. In some species, the setae can inflict painful stings. Some worms are equipped with chitinous jaws which they can use to give a painful bite. The worms with biting jaws include *Glycera dibranchiata.* Those worms having stinging setae include species of *Eurythöe* and *Hermodice.* Annelid worms are usually encountered when turning over rocks or coral boulders. *Glycera* ranges in coastal waters from North Carolina to northeast Canada. *Hermodice* and *Eurthöe* are found largely in tropical waters.

The chitinous jaws of *Glycera* are able to penetrate the skin and produce a painful sensation similar to that of a beesting. The wounded area may become hot and swollen, and this condition may remain for several days. The swelling may be followed by numbness and itching. Contact with the setae *Hermodice* or *Eurythöe* may produce swelling and numbness which may persist for several days.

Treatment and prevention. There is no specific treatment for *Glycera* bites. Immersing the hand of the victim in hot water for a period of about 30 minutes is the best treatment. Bristles can best be removed from the skin of the victim with the use of adhesive tape. Application of ammonia water or alcohol to the wound is useful in alleviating discomfort. Avoiding contact with marine worms is the best prevention. Divers should be particularly careful when reaching into crevices or under rocks.

Anemones, *like all Cnidarians, deliver venom by means of stinging cells called nematocysts.*

Sea Urchins

Sea urchins are free-living echinoderms, having a globular, egg-shaped, or flattened body. The viscera are enclosed within a hard shell, or test, formed by regularly arranged plates. Spines extend from tubercles on the test. Between the spines are three-jawed pedicellariae, pincerlike devices which inflict painful venomous bites in some species. The spines of some species are also poisonous. Among several hundred species of sea urchins the most dangerous are: the long-spined sea urchin (*Diadema* spp.), found in tropical seas; the venomous sea urchin (*Toxopneustes* spp.), found in the tropical Indo-Pacific region; and the *Asthenosoma ijimai,* which has no common name and is found in the Indo-Pacific region.

The venom apparatus of sea urchins consists of either the hollow, venom-filled spines or

Fire coral *has overgrown this harmless sea fan, making it painful to touch.*

the globiferous pedicellariae. Only one type of venom apparatus is present in a single species of sea urchin. The spines of most sea urchins are solid and have blunt, rounded tips. This type of spine is not a venom organ. However, some species have long, slender, hollow, sharp spines, which are extremely dangerous to handle. The acute tips and the spinules permit ready entrance of the spines deep into the flesh. But because of their extreme brittleness, they break off readily in the wound and are very difficult to withdraw. The spines of *Diadema* may reach a foot or more in length. Some of the spines of *Asthenosoma* are developed into special venom organs, each carrying a single large gland. The point of the spine is sharp and serves as a means of injecting the venom.

Pedicellariae are small, delicate, seizing organs found scattered among the spines of sea urchins. A special type of pedicellariae is called the globiferous, because of its globe-shaped head, and it serves as a venom organ. A sensory bristle is located on the inside of each valve. If the sensory bristle of the pedicellariae is touched, the small muscles at the base of a valve contract, thus closing the valve and injecting venom into the victim.

Penetration by the needle-sharp sea urchin spines may produce an immediate and intense burning sensation. The pain is soon followed by redness, swelling, and aching. Numbness and muscular paralysis have been reported. The sting from sea urchin pedicellariae may produce an immediate, intense, radiating pain, faintness, numbness, muscular paralysis, loss of speech, respiratory distress, and sometimes death.

Treatment and prevention. There are no specific antidotes, but soaking the affected part in hot water is usually helpful. The pedicellariae must be promptly removed from the wound, since they are usually detached from the parent animal and may continue to introduce venom into the wound. Sea urchin spines are extremely difficult to remove from the flesh because of their brittleness. The spines of some species are absorbed by the human body within 24 to 48 hours, but others must be removed surgically.

No sea urchin having elongate, needlelike spines should be handled. Although leather and canvas gloves, shoes, and flippers do not afford protection against spines, they can shield against pedicellariae. A diver working at night must be particularly careful.

Molluscs

Venomous molluscs consist of two major types: cone shells and octopuses.

Cone shells. Cone shells are named for their shape. There are about 400 species of cone shells, and all of them are equipped with a venom apparatus. Several of the tropical species can cause death. Some of the more dangerous species are the geographer cone, textile cone, tulip cone, and court cone, all found in the Indo-Pacific region.

The venom apparatus of cone shells consists of the venom bulb, venom duct, radular sheath, and radular teeth, which look like miniature spears. The entire complicated system lies in a cavity within the animal. It is believed that prior to stinging, the radular teeth, which are housed in the radular sheath, are released into the pharynx and thence to the proboscis. There they are grasped for thrusting into the flesh of the victim in the same fashion as an arrow is fired by an archer. The venom is probably forced under pressure by contraction of the venom bulb and duct into the radular sheath and then into the coiled radular teeth.

Stings produced by *Conus* are puncture wounds. Localized ischemia (lack of blood), cyanosis (blue color), numbness in the area about the wound, or a sharp stinging or burning sensation are usually the initial symptoms. The numbness may spread throughout the body. Paralysis, coma, and death by cardiac failure may result.

Treatment and prevention. There is no specific treatment or antivenin available. Symptomatic treatment is the same as that for venomous fish stings.

Live cone shells should be handled with care, and effort should be made to avoid contact with the soft parts of the animal.

Octopuses. The most dangerous octopuses from the standpoint of venom are some of the smaller species. The blue-spotted octopus *(Octopus maculosus)*, which is an Indo-Pacific species, has caused a number of fatalities in Australia.

The venom apparatus of the octopus includes the anterior and posterior salivary glands, the salivary ducts, the buccal mass, and the mandibles, or beak, which resembles a parrot's beak. The mouth is situated in the center of the anterior surface of the arms, surrounded by a circular lip which is fringed with fingerlike papillae. The chitinous jaws of the beak can bite vertically with great force, tearing the captured food which is held by the suckers before it is passed on to the rasping action of the radula. The venom is discharged from the anterior salivary glands into the pharynx by a pair of ducts.

Octopus bites usually consist of two small puncture wounds which are produced by the jaws. A burning or tingling sensation is generally the initial symptom. Bleeding may be profuse. Swelling, redness, and heat commonly develop in the area of the wound. The symptoms also include dryness of the mouth, difficulty in swallowing, vomiting, loss of muscular control, respiratory distress, and inability to speak. Death may result.

Treatment and prevention. There is no specific antidote. Octopus bites should be treated as any other venomous fish sting. Regardless of size, octopuses should be handled with gloves. If you have to kill an octopus, biting between the eyes is a sure method.

Fish

There are numerous species of fishes which are known to be venomous, but only a relatively few are commonly encountered by swimmers and divers.

Shiny sharks. Only one species of "horned" or "spiny" sharks is known to have stung men —the spiny dogfish *(Squalus acanthias)*, which ranges on both sides of the North Atlantic and North Pacific oceans.

Wounds are inflicted by the dorsal stings, which are adjacent to the front edge of the two dorsal fins. The venom gland appears as a glistening, whitish substance situated in a shallow groove on the back of the upper portion of each spine. When the spine enters the flesh of the victim, the venom gland is ruptured and the venom is released.

Symptoms are immediate, intense, stabbing pain, which may continue for hours. The affected part may redden and swell severely.

Treatment and prevention. Treat a shark sting as any other fish sting.

Care should be taken when removing the dogfish from a spear, hook, or net. The fish may give a sudden jerk and drive its sting deep into the flesh of the victim.

Stingrays. About 1500 stingray casualties are reported in the U.S. each year. Stingrays are divided into about seven families and a great many species. Common species include: diamond stingray (British Columbia to Central America); European stingray (northeastern Atlantic Ocean, Mediterranean Sea, and Indian Ocean); spotted eagle ray (tropical and warm temperate Atlantic, Red Sea, and Indo-Pacific); butterfly ray (Point Conception, California, south to Mazatlan, Mexico) California bat ray (Oregon to Magdalena Bay, Baja, California); and round stingray (Point Conception, California, south to Panama Bay).

Rays are common in tropical, subtropical, and warm temperate seas. They may be observed lying on top of the sand, or partially buried, with only their eyes, spiracles, and a portion of the tail exposed. Rays burrow into the sand and mud, and excavate the bottom with their pectoral fins, obtaining worms, molluscs, and crustaceans.

The venom apparatus of stingrays is an integral part of its tail appendage. It consists of the separate spine and an enveloping sheath of skin, together termed the sting. Stingrays usually possess only a single sting, but it is not unusual to find a specimen with two or more. The spine is made of a hard, bonelike material called vasodentine. Along either side of the spine are a series of sharp recurved teeth. Along either edge, on the underside of the spine, there is a deep groove.

*The **octopus** is a mollusc with a venomous bite and some can have a paralytic effect on man.*

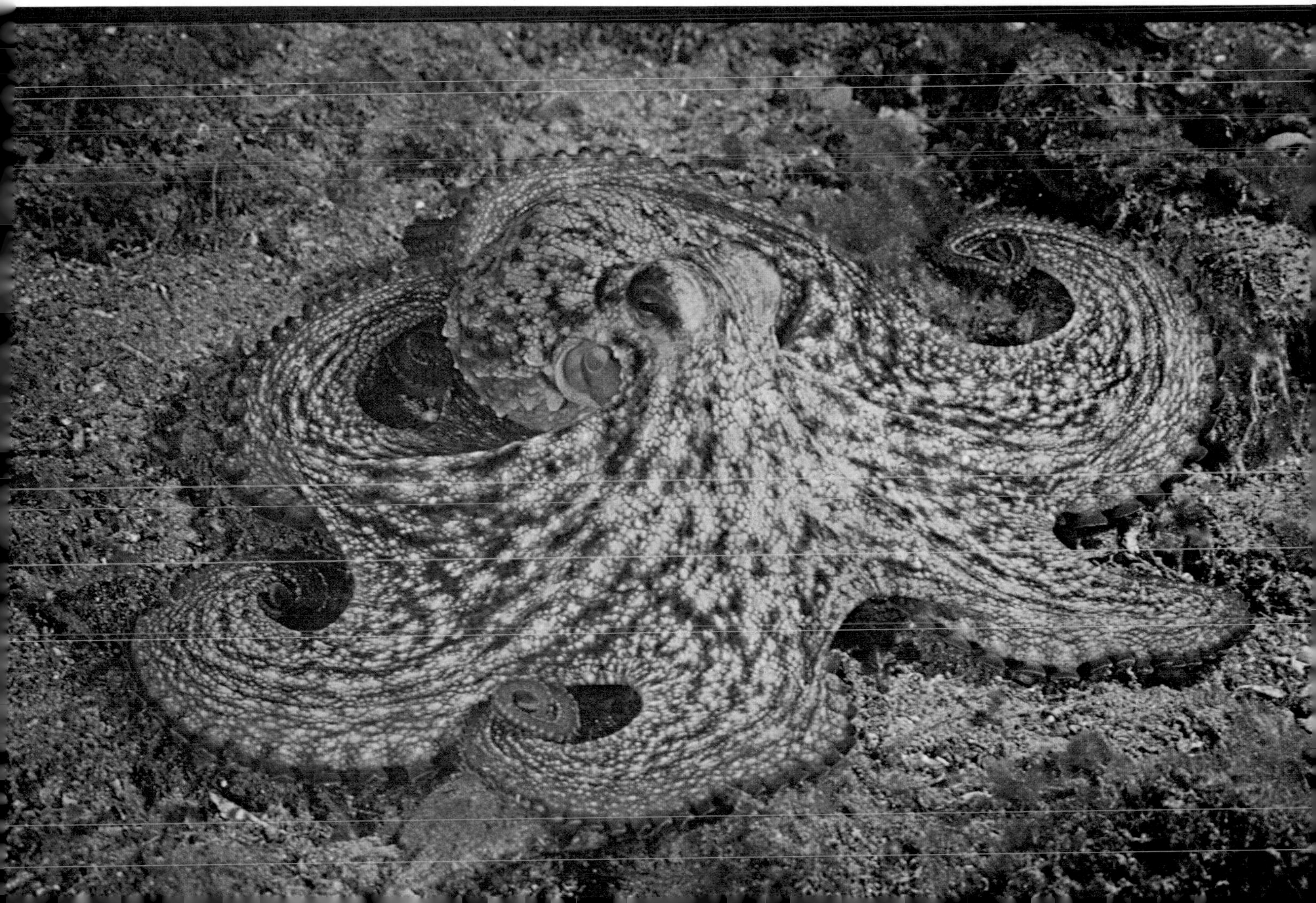

And in each of these grooves there is a strip of soft, spongy, grayish tissue, which produces the bulk of the venom. The grooves protect the soft delicate glandular tissue which lies within them.

Pain is the predominant symptom of a ray's sting, and it usually develops immediately or within a few minutes after the strike. The pain may be sharp, shooting, spasmodic, or throbbing. There may be a fall in blood pressure, vomiting, diarrhea, sweating, rapid heartbeat, muscular paralysis, and even death in rare cases. The venom acts directly on the heart and blood vessels and to some extent on the nervous system.

Treatment and prevention. There is no known specific antidote, and the injury should be treated as any other fish sting.

It should be remembered that stingrays commonly lie almost buried in the upper layer of a sandy or muddy bottom. Therefore they are a hazard to anyone wading in water inhabited by them. The body of the ray may be pinned down by the weight of the victim, thereby permitting the beast to arch its tail and make a successful strike. This danger can be eliminated by shuffling one's feet along the bottom or by using a stick to probe the bottom.

Catfish. Only a few of the approximately 1000 species of catfish live in salt water—the sea catfish (Cape Cod to Gulf of Mexico); the labyrinthic catfish (Indo-Pacific region, India); the catfish (Indo-Pacific region, India); and another sea catfish (Cape Cod to Brazil).

Venomous catfish have a single, sharp, stout spine immediately in front of the soft-rayed portion of the dorsal and pectoral fins. These spines are enveloped by a thin layer of skin, the integumentary sheath, which includes a series of venom-producing glandular cells. The spines of some species are also equipped with a series of sharp, recurved teeth which can severely lacerate the victim's flesh. The spines of the catfish are particularly dangerous because they can be locked into the extended position.

A catfish sting produces an almost instantaneous stinging, throbbing, or a scalding sensation which may be localized or may radiate up the affected limb. The area around the wound may become pale, surrounded by an area of redness and swelling. In severe cases numbness and gangrene may occur, so medical attention should be sought at once.

Treatment and prevention. There is no known specific antidote, and the injury should be treated as any other fish sting. Care should be taken in handling catfish because of their sharp rigid venomous fin spines.

Weeverfish. These are small marine fish which attain a length of 18 inches or less. But they are among the more venomous fish. Weevers live primarily in flat, sandy, or muddy bays. They are commonly seen burying themselves in the soft sand or mud, leaving only their head partially exposed. They may dart out rapidly and strike an object with unerring accuracy with their cheek spines. When a weever is provoked, its dorsal fin becomes erect instantly, and its gill covers expand. Because of its habit of concealment, aggressive attitude, and highly developed venom apparatus, the weever is a real danger to skin divers working in its habitat. Two of the four species of weeverfish are the great weever (Norway, British Isles, southward to the Mediterranean Sea; coasts of North Africa; Black Sea) and the lesser weever (North Sea, southward along the coast of Europe and Mediterranean Sea).

The venom apparatus of the weeverfish consists of the dorsal and opercular spines and their associated venom glands. The dorsal spines vary from five to seven in number.

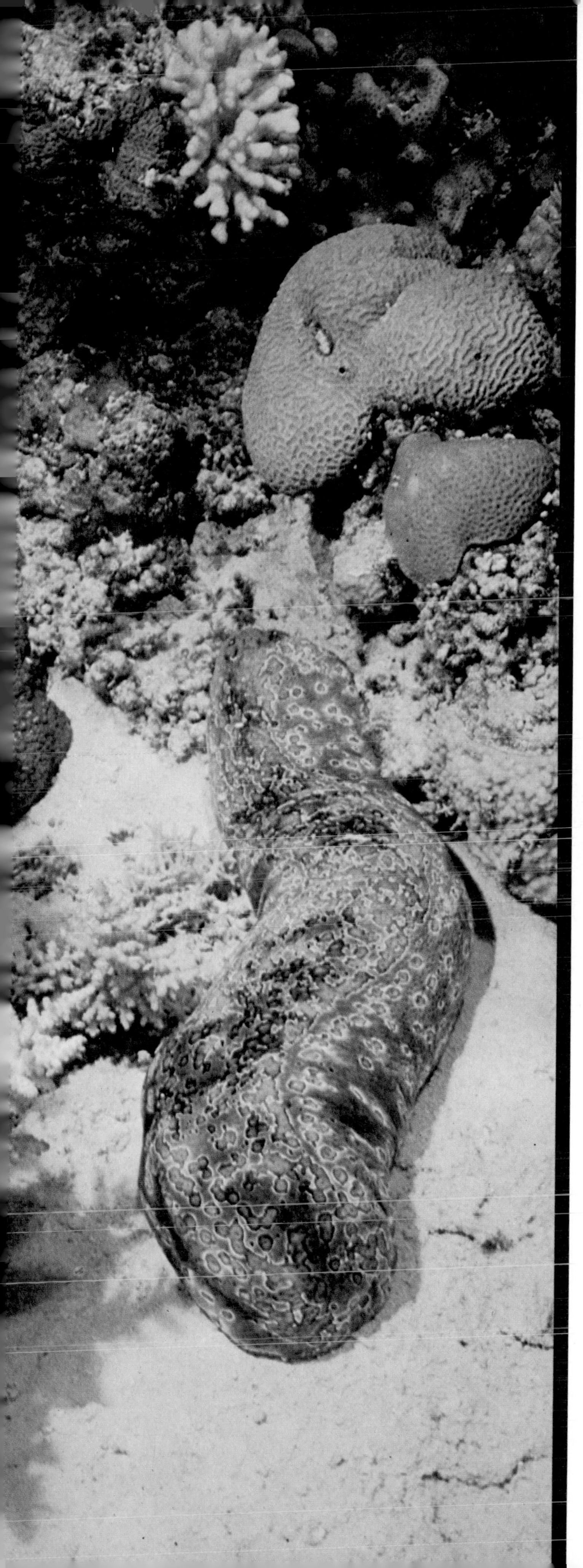

Each of the spines is enclosed in a thin-walled sheath of skin from which a needle-sharp tip protrudes. The gill covers are equipped with a daggerlike spine. Attached to the upper and lower margins of the spines are pear-shaped venom glands. The venom from weeverfish is similar to some snake venoms, in that it acts as both a nerve poison and a blood poison.

Weever stings produce instant burning, stabbing, or crushing pain, which is initially confined to the immediate area of the wound and then gradually spreads throughout the affected limb. The pain becomes progressively worse until it reaches an excruciating peak within about 30 minutes. The severity is such that the victim may scream, thrash about, and even lose consciousness. In some instances, even morphine fails to give relief. Numbness in the area of the wound may develop. At first, the skin about the wound is blanched, but it soon becomes reddened, hot, and swollen. In severe cases there may be nausea, vomiting, dizziness, sweating, cyanosis, joint aches, loss of speech, difficulty in breathing, gangrene, and even death.

Treatment and prevention. No specific antidotes are generally available, and the injury should be treated as any other fish sting.

Weeverfish stings are commonly encountered while wading or swimming along sandy coastal areas. Persons wading in waters where weevers abound should wear adequate footwear. Skin divers should attempt to avoid antagonizing these little fish since they are easily provoked into stinging. No one should ever attempt to handle a living weeverfish. Even when dead their spines and venom apparatus may be capable of inflicting a painful wound.

Considered a delicacy in the Orient, some species of ***sea cucumber*** *have body fluids toxic to man.*

Scorpionfish. These fish are widely distributed throughout all tropical and temperate seas. A few species are also found in arctic waters. Venomous scorpionfish are divided into three main groups on the basis of the structure of their venom organs—zebrafish *(Pterois),* scorpionfish proper *(Scorpaena),* and stonefish *(Synanceja).*

Zebrafish are among the most beautiful and ornate of coral reef fish. They are generally found in shallow water, hovering about in a crevice or at times swimming unconcernedly in the open. They are also called turkeyfish because of their interesting habit of slowly swimming about, spreading their fanlike pectorals and lacy dorsal fins like a turkey gobbler displaying its plumes. But succumbing to the temptation to reach out and grab one of these fish results in an extremely painful experience. Hidden beneath the lace are the fish's needle-sharp fin stings.

The venom apparatus of the zebrafish consists of thirteen dorsal spines, three anal spines, two pelvic spines, and their associated venom glands. For the most part the spines are long, straight, and slender; they are camouflaged in delicate, lacy-appearing fins. Located on the front side of each spine are the glandular grooves. Open on either side, they are deep channels extending the entire length of the shaft. Situated within these grooves are the venom glands, which are enveloped by a thin integumentary sheath.

The scorpionfish proper are generally bottom-dwellers, found in bays, along sandy beaches, rocky coastlines, or coral reefs, from the intertidal zone to depths of 50 fathoms or more. Their habit of concealing themselves, together with their protective coloration which blends them almost perfectly into their surrounding environment, makes them difficult to see. When they are removed from the water, they possess the defensive habit of erecting their spinous dorsal fin and flaring out their gill covers and pelvic and anal fins. All of these are armed except the pectoral fins. The scorpionfish have a variable number of dorsal spines, but most often twelve; three anal spines; two pelvic spines; and their associated venom glands. The spines are shorter and heavier than those found in the zebrafish. The venom glands lie along the glandular grooves, but are limited to the distal half of the spine.

Stonefish are generally shallow-water dwellers and are commonly found in tide pools and shoal reef areas. *Synanceja* has the habit of lying motionless in coral crevices, under rocks, in holes, or buried in sand and mud. They appear to be fearless and completely disinterested in intruders. The stonefish usually has thirteen dorsal spines, three anal spines, two pelvic spines, and their associated venom glands. The venom organs are unique among this group because of the short, heavy spines and enlarged venom glands, which are covered by a thick layer of warty skin.

The symptoms produced by the sting of various species of scorpionfish are essentially the same; they vary in degree, rather than in quality. The pain is usually sharp, shooting, or throbbing, and radiates from the affected part. The area around the wound becomes blanched and later cyanotic. The pain from a stonefish sting can be extreme, causing the victim to lose consciousness. The area about the wound becomes reddened and swollen. Later the wound may ulcerate. Serious cases may exhibit nausea, vomiting, swelling of the lymph nodes, joint aches, respiratory distress, convulsions, and death. Scorpionfish venoms appear to affect the heart and blood vessels, as well as the nerves and muscles.

Treatment and prevention. A special antivenin has been developed by the Commonwealth Serum Laboratories in Melbourne,

Australia, for use on stonefish stings. If the antivenin is not available, the injury should be treated as any other fish sting.

Zebrafish should never be handled, and great caution should be used in removing scorpionfish from hook or nets. Stonefish are especially dangerous because they are so well camouflaged. Placing one's hands in crevices or in holes that may be inhabited by these fish should be avoided.

Toadfish. These are small bottom fish which inhabit the warmer waters in the Red Sea, the Indian Ocean, the Mediterranean and nearby Atlantic coasts, and the Pacific coast of Central America. Toadfish have broad, depressed heads and large mouths and are somewhat repulsive in appearance. They hide in crevices, under rocks and debris, and among seaweed; or they may lie almost completely buried under sand or mud.

***Stingrays** can inflict painful wounds with the barbed, venomous spine at the base of their tail.*

The venom apparatus of toadfish consists of two dorsal fin spines, two gill cover spines, and their associated venom glands. The dorsal spines are slender and hollow, slightly curved, and terminate in sharp, needlelike points. At the base and tip of each spine is an opening through which the venom passes. The base of each dorsal spine is surrounded by a glandular mass producing the venom. The spine on each gill cover is made from a slender extension of the operculum, a hollow bone which curves slightly and terminates in a sharp tip. With the exception of the outer tip, the entire gill spine is encased in a glistening, whitish, pear-shaped mass. This mass is the venom gland. The gland empties into the base of the hollow gill spine which serves as a duct. When a bather or diver contacts the spine, the venom is delivered.

The pain from toadfish wounds develops rapidly, is radiating and intense. Some have described the pain as similar to a scorpion sting. The pain is soon followed by swelling, redness, and heat.

Treatment and prevention. There are no specific antidotes, and the injury should be treated as any other fish sting.

Persons wading in waters inhabited by toadfish should take the precaution of shuffling their feet through the mud to avoid stepping on the fish. Removal of toadfish from a hook or from nets should be done with care.

Stargazers. These are small, carnivorous, bottom-dwelling marine fish. They have a cuboid head, an almost vertical mouth with fringed lips, and eyes on the flat upper surface of the head. Stargazers spend a large part of their time buried in the mud or sand with only their eyes and a portion of the mouth protruding, waiting for prey. Some representatives of stargazers are found in the seas off southern Japan, southern Korea, China, the Philippines, and Singapore; in the Indo-Pacific region; and in the eastern Atlantic and Mediterranean Sea.

The venom apparatus of stargazers consists of two shoulder spines, one on each side, protruding through a sheath of skin. The venom glands are attached to these spines. Wounds from stargazers are said to be painful, but little is known about the symptoms.

Treatment and prevention. There are no specific antidotes, and the injury should be treated as any other fish sting. Stargazers should be handled with extreme care to avoid being jabbed by the shoulder spines.

Rabbitfish. These spiny-rayed fish closely resemble the surgeonfish. They differ from all other fish in that the first and last rays of their pelvic fins are spinous. Rabbitfish are of moderate size, usually valued as food, and abound about rocks and reefs from the Red Sea to Polynesia. Most of them are members of the genus *Siganus*.

The rabbitfish's venom apparatus consists of thirteen dorsal, four pelvic, and seven anal spines, and their associated venom glands. A groove extends along both sides of the midline of the spine for almost its entire length. These grooves are generally deep and contain the venom glands, which are located in the outer one-third of the spine near the tip.

The symptoms resulting from rabbitfish stings resemble those of scorpionfish stings.

Treatment and prevention. These injuries should be treated as any other fish sting.

Treatment of venomous fish stings. Treatment should be directed toward three objectives: alleviating pain, combating effects of the venom, and preventing secondary infection. Pain results from the trauma produced by the fish spine and venom and from the introduction of slime and other irritating foreign substances into the wound. In the case of stingray and catfish stings the retrorse barbs of the spine may produce severe lacerations with considerable trauma to the soft tissues. Wounds of this type should be promptly irrigated or washed out with warm

Puffers, *universally regarded as poisonous, produce one of the most lethal fish toxins known.*

salt water or sterile saline solution if available. Fish stings of the puncture-wound variety are usually small in size and removal of the poison is more difficult. It may be necessary to make a small incision across the wound and then apply immediate suction and irrigation. However, fish do not inject their venom as do venomous snakes, so suction is not very effective.

The most effective treatment of fish stings is to soak the injured member in hot water for a period of 30 minutes to one hour. The water should be maintained at as high a temperature as the patient can tolerate without injury, and the treatment should be started as soon as possible. If the wound is on the face or body, hot moist compresses should be employed. The addition of magnesium sulfate or epsom salts to the water may be useful. This is where first aid ends. Whenever possible, a doctor should be called at once, or the victim should be taken to a hospital. Under medical supervision only, infiltration of the wound area with procaine or xylocaine may be necessary. If local measures fail to prove satisfactory, intramuscular or intravenous demerol may be required. Following the soaking procedure, debridement and further cleansing of the wound may be desirable. Lacerated wounds should be closed with dermal sutures. If the wound is large, a small drain should be left in it for a day or two. The injured area should be covered with an antiseptic and a sterile dressing.

*The **surgeonfish** has a razor-sharp venomous spine on the side of its tail that wards off attackers.*

Prompt institution of the recommended measures usually eliminates the necessity of antibiotic therapy. But if treatment has been delayed, administration of antibiotics may be required. If the victim has previously received a tetanus immunization, tetanus toxoid should be given (0.5 cc.). If the patient has not received a previous tetanus immunization, he should be given a passive immunization with 250 units of human tetanus-immune globulin administered intramuscularly as soon as possible.

The primary shock that immediately follows the sting generally responds to simple supportive measures. However, secondary shock resulting from the action of stingray venom on the cardiovascular system requires immediate and vigorous therapy. Treatment should be directed toward maintaining cardiovascular tone and preventing further complications. Respiratory stimulants may be required. The services of a physician are needed in treatment beyond initial first aid.

Sea Snakes

There are about 52 species of sea snakes, all of which are inhabitants of the tropical Pacific and Indian oceans. All are marine except a single freshwater species *(Hydrophis semperi)*, which is found in the freshwater Lake Taal in the Philippines. Sea snakes are characteristically residents of sheltered coastal waters and are particularly fond of river mouths. *Pelamis platurus* is the most widely distributed species, ranging from the west coast of Central America westward to the east coast of Africa. The remainder of sea snakes are primarily Indo-Pacific forms. With their compressed, oarlike tails, sea snakes are well adapted for locomotion in their marine environment, swimming by lateral undulatory movements of the body. They have a remarkable ability to move forward or backward in the water with equal rapidity, but they are awkward on land. Although they depend upon air for their respiration, sea snakes are able to remain submerged for hours. They capture their food underwater, and their diet consists almost entirely of fish. A considerable portion of their time is spent feeding on or near the bottom around rocks and in crevices. There they capture eels and other small fishes, which they promptly kill with a vigorous bite of their venomous fangs. Generally speaking, sea snakes tend to have a rather docile disposition, but during the mating season some species can become quite aggressive. In a series of 120 cases of sea snake bites, it was found that most of the attacks occurred while handling nets, sorting fish, wading, washing, or accidentally stepping on the snake. It should be kept in mind that the venom of sea snakes is extremely potent and capable of producing death.

Sea snakes inflict their wound with fangs, which are quite small, but similar to those of the cobra. Compared to that of other venomous snakes, the dentition of sea snakes is relatively feeble but is nevertheless fully developed for venom conduction. The venom glands are situated one on either side behind and below the eye, in front of the tympanic bones. Most sea snakes have two fangs on each side, but some have only one on each side. The venom duct discharges its contents at the base of the fang.

Symptoms caused by the bite of a sea snake tend to develop rather slowly. It may take 20 minutes to several hours before clinical evidence appears. Aside from the initial prick, there is no pain or reaction at the site of the bite. The victim sometimes fails to associate the bite with the subsequent illness. Symptoms may include mild euphoria, aching, anxiety, sensation of thickening of the tongue, and stiffness of the muscles. Paralysis may follow; it may become generalized, beginning with the legs and gradually moving up the body. Drooping of the eyelids is an early sign. Speaking and swallowing becomes increasingly difficult. Thirst, burning, and dryness of the throat may be present. Other symptoms are muscle twitching, spasms, ocular and facial paralysis, clammy skin, cyanosis, convulsions, respiratory distress, and loss of consciousness. Death results from respiratory failure.

*Graceful **zebrafish** are unafraid of divers venomous spines provide security against being molested.*

Treatment and prevention. The routine incision and suction method that is generally used in the treatment of snake bite is not recommended. The victim should avoid all possible exertion. A tourniquet should be applied to the thigh in leg bites or to the arm above the elbow in bites of the hand or wrist. The tourniquet should be released every 30 minutes. Keep in mind that you cannot cut off the blood supply to an appendage, then just leave it. The victim should be carried to the nearest first-aid station or hospital as soon as possible. Do not make him walk. The snake should be captured and sent to the hospital for identification. This is very important because the snake may be a harmless water snake. Antivenin therapy should be administered promptly upon arrival at the first-aid station. If available, sea snake antivenin should be used. If not available, a polyvalent antiserum containing a krait (Elapidae) fraction should be used. The services of a physician are required.

The bulk of recorded attacks have occurred among fishermen working with their nets in the vicinity of river mouths. In most instances, the attacks have resulted because of provocation of the snake. Individuals wading and divers working around rocky crevices, piers, and old tree roots inhabited by sea snakes should be aware of the danger. Despite the generally docile nature of sea snakes, except possibly during the mating season, an attempt should be made to avoid handling or coming in contact with them. If bitten try to exercise as little as possible and seek immediate medical help.

Venomous spines and pedicellariae (pincerlike organs) serve as protection for some ***sea urchins.***

Chapter VI. Filing Sea Life (Taxonomy)

The first practical and still valid system of ordering the chaotic tangle of living organisms was described by Carolus Linnaeus (1707–78) in a mid-eighteenth-century publication entitled *Systema naturae*. The Linnaean system of classification involves a hierarchical (stepwise) arrangement of groupings as follows:

KINGDOM
  PHYLUM
    Class
      Order
        Family
          Genus
            species

This system is still the best in taxonomy—that is, the study and process of the classification of things—and is in use in almost the original form today. The International Committee of Zoological Nomenclature has been established to oversee the application of Linnaean taxonomy and the standardization of taxonomic names and hierarchies for newly described creatures.

Science cannot progress without a system of identification to make communication possible: in order to know something it must first be classified. In other words, any object or being must be described, classified, and viewed in its relationship to other objects or animals. Common practice in modern taxonomy is to refer to an animal or plant by the generic and specific names—thus, *Canis familiaris,* the dog—and these names are italicized in common text. Following the Linnaean system, one looks to the next higher category for the near relations. If *familiaris* is the species of dogs (and the species concept is a complex study in itself), next higher is the genus *Canis,* which also includes wolves and jackals. This genus is in the family that also includes bears and other similar creatures. By this process the degree of relationship to every other living form can be traced.

In modern practice it is recognized that these classifications are quite artificial. Animals and plants do not produce equal-level groupings, and thus qualifications are built into the system. There are sub-, infra-, and super-prefixes for each hierarchical level, and consequently we find categories called the superfamily, infraclass, and suborder. Even with this tremendous flexibility, there is still something forced in the practice of taxonomy, but scientists must live with this as a better system has never been proposed.

No two taxonomists are likely to produce the same result, even with the standardization that the international commission provides. In the system presented here we have extracted the results of worldwide research. In many cases this taxonomy goes down to the ordinal (order) level: further breakdown of the list would require many volumes. The emphasis is on life from the sea. The arrangement is pseudophylogenetic (from simple to complex), but there are many cases where the exact relative levels of complexity are unknown. Much of this taxonomy is based on hard data, much on educated guesswork. As a final note, it is interesting to observe that the creatures that appealed most (such as the birds and insects) to scientists and nature buffs tend to have the longest lists of subdivisions. This is not a reflection of nature, but rather one of man.

To know a creature, the scientist must be able to classify it and thus separate it out of the bewildering number of organisms. The ***sea anemone*** *is an animal of the Phylum Cnidaria.*

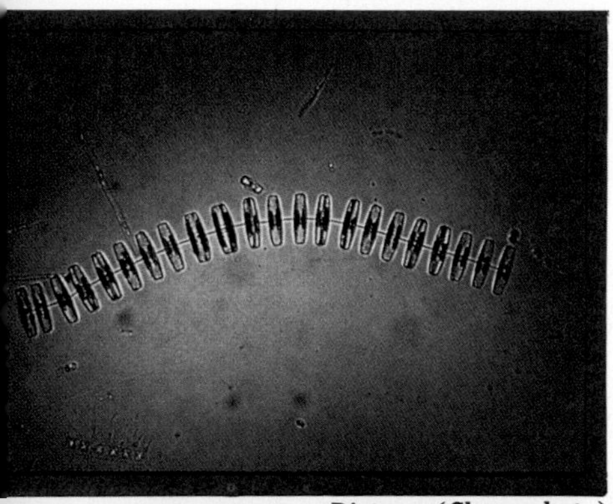

Diatoms (Chrysophyta)

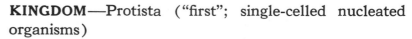

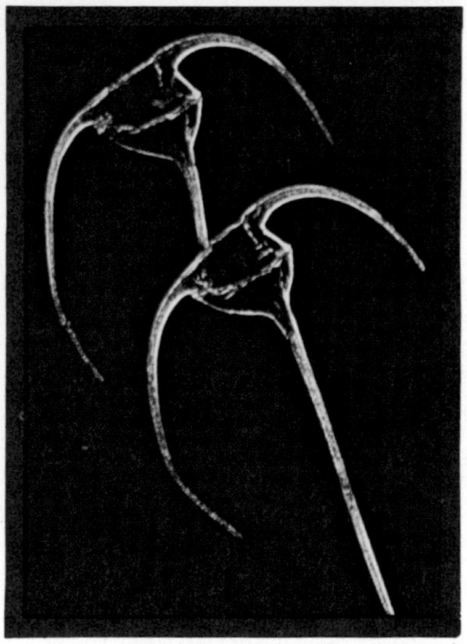

Dinoflagellates (Pyrrophyta)

Taxonomy (with an emphasis on sea-life)

KINGDOM—Monera (organisms without nuclear membranes)

Simplest of all organisms, some are photosynthetic. All procaryotic (lacking nuclear membranes). Chromosomes and other genetic material clumped together in aggregates throughout the cytoplasm (the basic substance of the cell). Cell walls present; some forms flagellate.

PHYLUM—Schizophyta ("divided-plant"; bacteria)

Unicellular forms with a rigid cell wall; surrounded by a capsule in some forms. Photosynthesis occurs in some heterotrophs (forms that depend on others for food production). Reproduction by binary fission. Common forms are coccoid, spirochaete, and coliform bacteria.

PHYLUM—Cyanophyta (blue-green algae)

May form colonies or exist as independent cells. Individual cells are spherical or cylindrical. None possess flagella. Cell walls composed of cellulose, all cells contain the blue pigment phycocyanin. Reproduction by fission or asexually by endospores (a spore that forms within the cells). Common forms: *Trichodesmium, Nostoc.*

KINGDOM—Protista ("first"; single-celled nucleated organisms)

Unicellular, some in colonies. Both autotrophic (forms that make their own food) and heterotrophic forms in this kingdom; the ancestors of all higher plants and animals. Most of the phytoplankton are protists, as well as the microscopic zooplankton.

PHYLUM—Protophyta ("first-plants")

Both heterotrophic and autotrophic; some forms possess cell walls of hemicellulose or silica, others lack cell wall. Motile species have flagella. Most forms in this phylum microscopic, some bioluminescent.

SUBPHYLUM—Chrysophyta (golden algae and diatoms)

SUBPHYLUM—Xanthophyta (yellow-green algae)

SUBPHYLUM—Pyrrophyta (dinoflagellates and cryptomonads)
SUBPHYLUM—Hyphochytridiomycota (hyphochytrids)
SUBPHYLUM—Plasmodiophoromycota (funguslike protists)
SUBPHYLUM—Euglenophyta *(Euglena)*

PHYLUM—Protozoa ("first -animals")

Single-celled protists with generally "animal" characteristics: no cell wall, no chloroplasts; sexual and asexual forms of reproduction throughout the phylum. Some have shells, most are microscopic.

SUBPHYLUM—Mastigophora (trypanosomes, flagellated forms)
SUBPHYLUM—Sarcodina (amoeboid forms)
Class—Actinopodea (includes radiolaria)
Class—Rhizopodea (true amoeboids)
Order—Amoebida *(Amoeba)*
Order—Testacida (freshwater shelled forms)
Order—Foraminifera (marine and freshwater forms)
SUBPHYLUM—Sporozoa (spore-forming parasites)
SUBPHYLUM—Ciliata (ciliated forms, *Paramecium)*

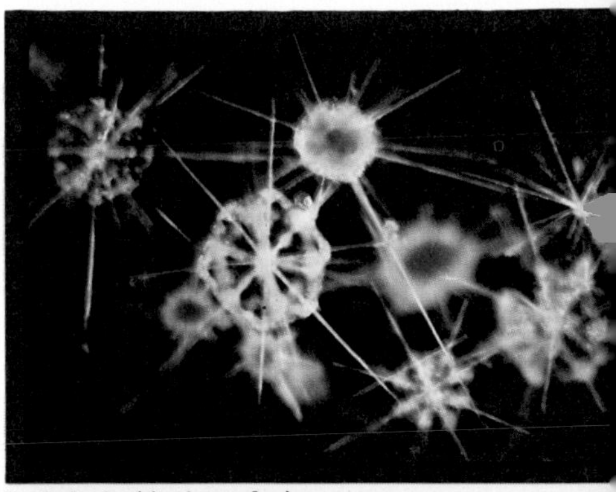
Radiolaria (Actinopodea)

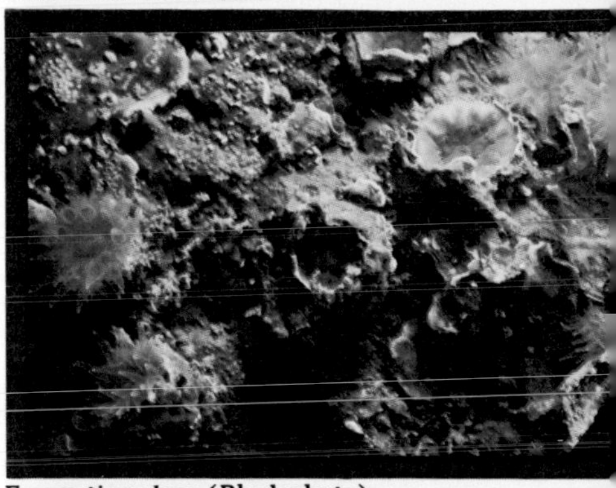
Encrusting algae (Rhodophyta)

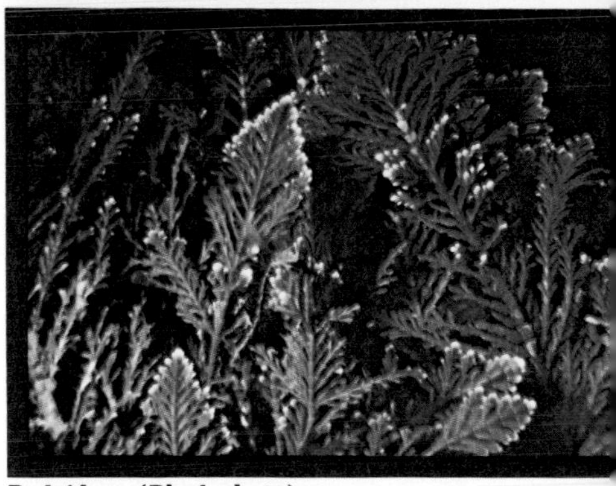
Red Algae (Rhodophyta)

KINGDOM—Plantae (true plants)

Mostly multicellular organisms with specialized photosynthetic structures. (Nonphotosynthetic fungi included because of the close structural relationships to other plants.) Cell differentiation and complex structural organization. Rigid cell walls usually composed of cellulose. Reproduction by sexual or asexual means. Botanists use "Division" in place of "Phylum" for the higher plants.

DIVISION—Rhodophyta (red algae)

Majority of species marine. Distinguished by presence of red pigment, phycoerythrin. All nonmotile, including their gametes; complex reproductive cycles. Most of the 2500 species are filamentous and small. Several form lime deposits that contribute to coral reef building.

Common representatives: Laver, Irish moss, dulse, coralline algae, *Dasya, Gelidium.*

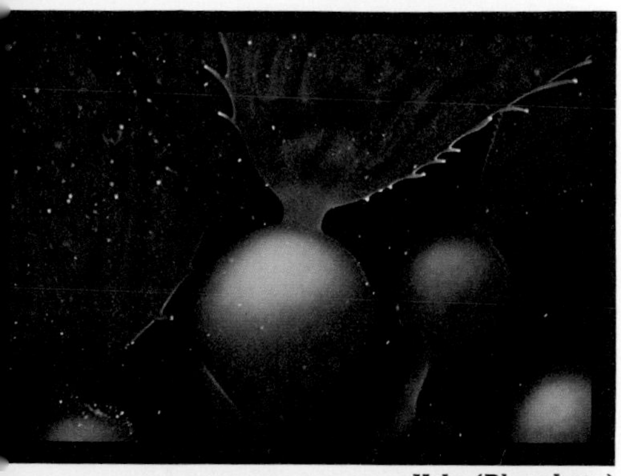

Kelp (Phaephyta)

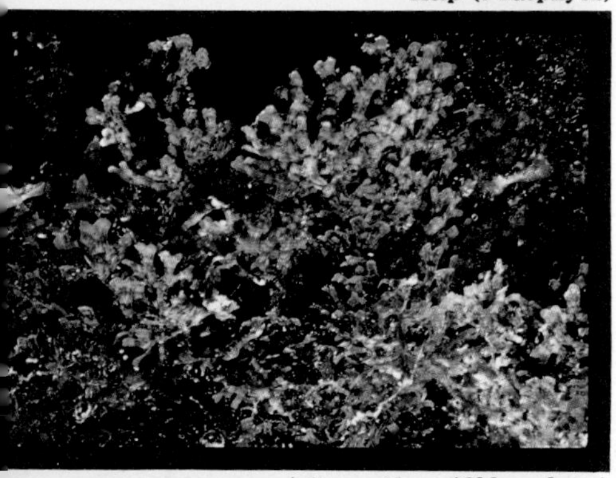

Green Algae (Chlorophyta)

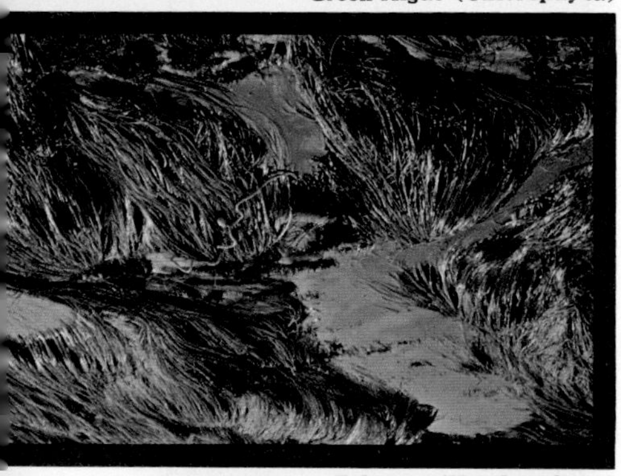

Eelgrass (Tracheophyta)

Vase sponge (Demospongiae)

DIVISION—Phaeophyta (brown algae)

All multicellular. Largest and most complex marine algae. Primarily a marine phylum containing over 1000 species. Reproduction generally sexual with alternating haploid and diploid generations. A rootlike organ, the hold fast, present in most. Distribution: arctic to temperate zones.

Common forms: *Fucus* (rockweed), *Sargassum* (sargasso weed), *Macrocystis* (kelp).

DIVISION—Chlorophyta (green algae)

The most numerous and diversified of the algal phyla, numbering 5700 species. Contains both marine and freshwater forms, some planktonic unicellular, others large and complex. A few are calcareous. The ancestors of land plants are from this phylum. Common forms: *Ulva* (sea lettuce), *Chlamydomonas, Halimeda.*

DIVISION—Bryophyta (mosses and liverworts)

All terrestrial, characterized by lack of vascular tissue. Possess rhizoids rather than true roots. Reproduction sexual, all thought to have evolved from filamentous green algae. Common forms: *Polytrichum,* peat moss.

DIVISION—Fungi (true fungi)

Possibly a separate kingdom; plants lacking photosynthetic abilities, yet absorbing substances through cell walls as plants do. The fungi probably diverged from the plants and animals early in life history. The group includes slime molds, yeasts, the fungal part of lichens, mushrooms, and toadstools.

DIVISION—Tracheophyta (vascular plants)

All plants possessing vascular tissue, true roots, stems, and leaves. Flowers, fruits, and seeds generally characters of the group called angiosperms. Includes trees, shrubs, flowering plants, ferns. Common marine genera: *Zostera* (eelgras), *Thalassia* (turtle grass), and *Spartina.*

KINGDOM—Animalia (all multicelled animals)

Forms which are heterotrophic, generally mobile, generally lacking starches (such as cellulose). Body plan on a tissue or organ level, reproduction may be sexual, asexual, or both within the same organism.

PHYLUM—Mesozoa ("middle-animals")

Parasitic animals, usually hosted by molluscs and other invertebrates. Their true phylogenetic position is obscure; they may be degenerate flatworms; body consists of a single layer of cells, a few interior reproductive cells.

PHYLUM—Porifera ("pore-bearing"; the sponges)

Most primitive metazoans, body radial or asymmetrical, usually hollow, walls perforated, interior lined with flagellated choanocyte cells. Almost all are marine. Horny sponges contain a unique material, spongin.

- *Class*—Calcarea (chalk sponges)
- *Class*—Hexactinellida (glass sponges)
- *Class*—Demospongiae (horny sponges)
- *Class*—Archaeocyathida (extinct, affinity uncertain)
- *Class*—Receptaculitida (extinct, affinity uncertain)

PHYLUM—Cnidaria or Coelenterata ("cnid" = nettle, "coel" = cavity; jellyfish, anemones, corals)

Mostly marine animals usually bearing tentacles with nematocysts (stinging cells). Digestive cavity with one opening. May have a medusoid and/or a polypoid stage, solitary or colonial. This group is close to the stem of the higher animal radiation, and reproduces via planula larvae, or asexually.

- *Class*—Hydrozoa (with hydroid and polypoid stages)
  - Order—Hydroidea (hydroids)
  - Order—Milleporina (stinging corals)
  - Order—Siphanophora (Portuguese man-of-war)
  - Order—Stylasterina (like Milliporina)
  - Order—Trachylina (medusoid forms)
- *Class*—Scyphozoa (free-swimming medusoids)
  - *Subclass*—Scyphomedusae (true jellyfish)
  - *Subclass*—Conulata (extinct, affinity uncertain)
- *Class*—Anthozoa (exclusively polypoid)
  - *Subclass*—Alcyonaria (octocorallia, eight branched tentacles)
    - Order—Stolonifera (organ pipe corals)
    - Order—Telestacea (resembles gorgonians)
    - Order—Alcyonacea (soft corals)
    - Order—Coenothecalia (blue corals)
    - Order—Gorgonarea (gorgonians, sea fans, red coral)
    - Order—Pennatulacea (sea pens, sea pansies)

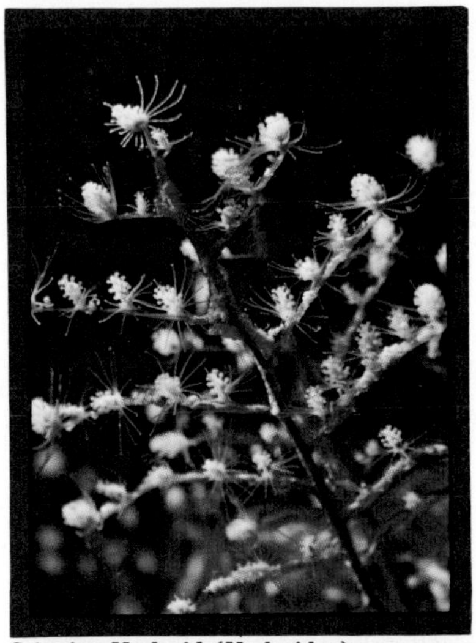

Stinging Hydroid (Hydroidea)

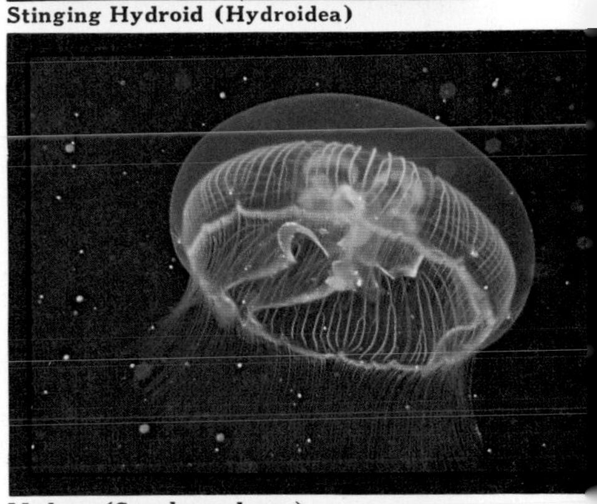

Medusa (Scyphomedusae)

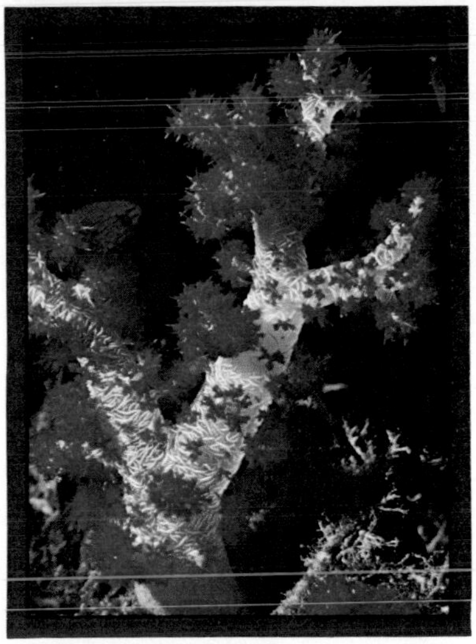

Soft Coral (Alcyonacea)

Mushroom Coral (Madreporina)

Tube Anemone (Actinaria)

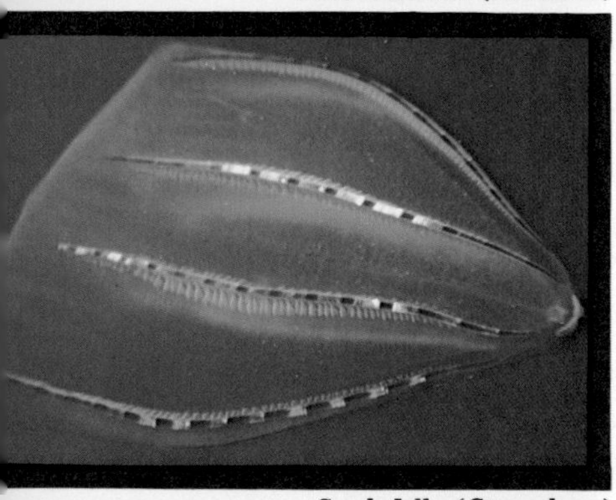
Comb Jelly (Ctenophora)

Flatworm (Turbellaria)

Subclass—Zoantharia (hexacorals, six or more unbranched tentacles)
- Order—Rugosa (fossil large forms)
- Order—Tabulata (fossil colonial forms)
- Order—Actinaria (sea anemones)
- Order—Madreporina (stony corals)
- Order—Zoanthidea (epizoic polyps, like small anemones)
- Order—Antipatharia (black corals)
- Order—Ceriantharia (cerianthid or burrowing anemones)

PHYLUM—Ctenophora ("comb-bearer"; comb jellies)

Exclusively marine, hermaphroditic, biradially symmetric animals that resemble jellyfish. Stinging cells are absent, locomotion is by eight comb plates, forms with and without tentacles. When present tentacles are retractable within internal chambers. All develop via a cydippid larva. Sizes range from one to six inches long.

PHYLUM—Platyhelminthes ("flat-worms")

Bilaterally symmetrical animals, flattened dorsoventrally (top-bottom). Two-thirds of the species are parasitic, all lack circulatory systems, anuses, most hermaphroditic. Turbellarians live in freshwater or may be marine.
- *Class*—Turbellaria (planarians, marine flatworms)
- *Class*—Trematoda (flukes)
- *Class*—Cestoidea (tapeworms)

PHYLUM—Nemertina (from *Nemertes,* "sea nymph"; ribbon worms)

Mostly marine infaunal forms, some pelagic. Possess circulatory system, anus, extensible proboscis. Distinct sexes, one species of *Lineus* may be 100 feet long.

PHYLUM—Aschelminthes (roundworms)

Probably an artificial assemblage of wormlike animals with external cuticles. Marine and nonmarine, free-living and parasitic forms. Sexes usually separate but the reproductive system is usually very simple.
- *Class*—Rotifera (rotifers)
- *Class*—Nematoda (nematodes)
- *Class*—Priapulida (rare marine forms)
- *Class*—Nematomorpha (hairworms)
- *Class*—Kinorhyncha (microscopic marine forms)

PHYLUM—Acanthocephala ("thorny-headed"; worms)

Endoparasites, inhabiting birds, mammals, and fish. Intermediate hosts are arthropods. All bear a probiscis with spines, many of the systems are degenerate from parasitic habit.

PHYLUM—Entoprocta ("internal-anus")

A small group (60 species) of marine (except one species) stalked animals, most microscopic. They are filter feeders, and the only nonwormlike pseudocoelomate (false cavity) forms. Mouth and anus within a circle of tentacles, reproduction via a unique form of larva.

PHYLUM—Phoronida ("nest-bearing")

Marine lophophorate (the food-gathering arm), tube-dwelling wormlike animals. Most small (under about one-half inch), some more than a foot long.

PHYLUM—Ectoprocta ("outside-anus"; bryozoans)

Mostly marine "moss animals," always in colonies, sessile. Individuals are small, the colonies usually encrusting. Colonies expand by budding, new colonies form from larvae. Very important in fossil record.

Class—Gymnolaemata (marine)
Class—Phylactolaemata (freshwater)

PHYLUM—Brachiopoda ("arm-foot"; lampshells)

Marine lophophorates with bivalved shells, usually stalked. Develop from unique free-swimming larva; the shells are bilaterally symmetrical within each valve. Very important as fossils, much rarer today.

Class—Inarticulata (simple forms, shells lack hinge teeth, *Lingula*)
Class—Articulata (shells with hinge teeth)

PHYLUM—Mollusca (from Latin *molluscus,* "soft")

Second-largest phylum of metazoans (higher animals). Bear gills, well-developed head (usually), chambered heart, eyes, sexes separate or hermaphroditic. Body plan bilateral, most representatives shelled. Larval development from trochophores often followed by veligers. Marine, freshwater, terrestrial.

Class—Amphineuria ("both-nerve"; primitive forms)
Order—Aplacophora (wormlike, without shells)

Phoronid (Phoronida)

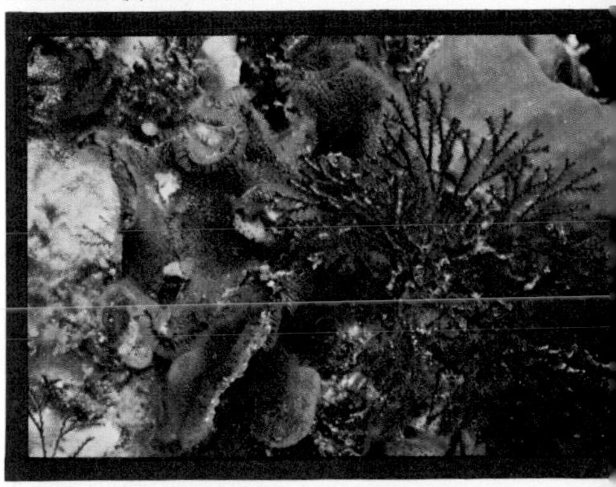

Bryozoan (Ectoprocta)

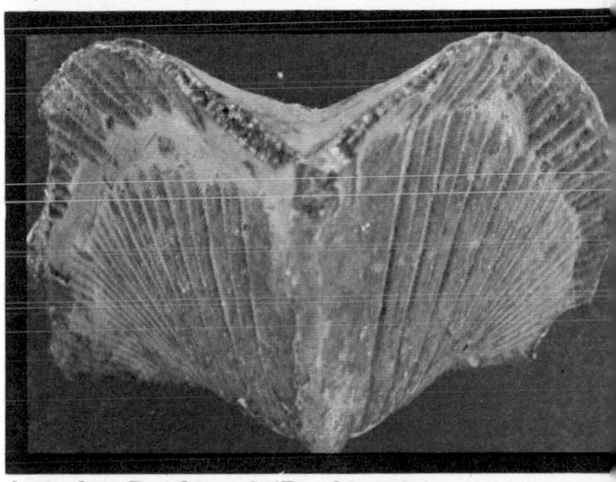

Articulate Brachiopod (Brachiopoda)

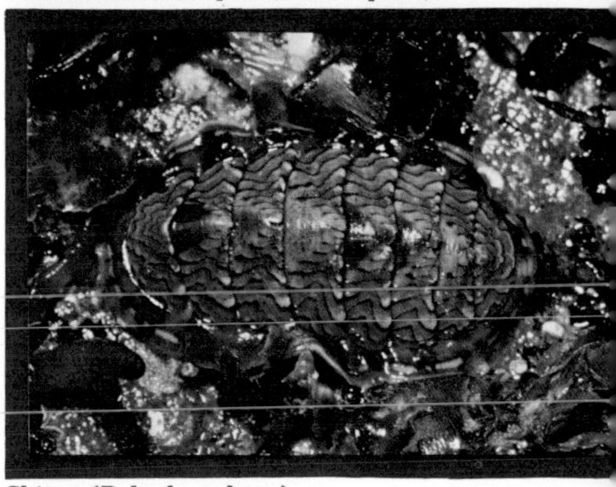

Chiton (Polyplacophora)

Sea Slug (Nudibranchia)

Snail (Neogastropoda)

File shell (Filibranchia)

Nautilus (Cephalopoda)

Order—Monoplacophora (*Neopilina*)
Order—Polyplacophora (chitons)
Class—Gastropoda ("stomach-foot"; snails)
Subclass—Prosobranchia (true snails)
Order—Aspidobranchia (limpets)
Order—Pectinobranchia (periwinkle, cowries)
Order—Neogastropoda (whelks, conchs, *Murex*)
Order—Archaeogastropoda (extinct)
Subclass—Opisthobranchia (reduced or absent shells)
Order—Tectibranchia (sea hares, pteropods)
Order—Nudibranchia (sea slugs)
Class—Scaphopoda ("ship-foot"; tusk shells)
Class—Pelecypoda ("hatchet-foot"; bivalves)
Subclass—Protobranchia (simple forms, *Nucula*)
Subclass—Filibranchia (mussels, pectens)
Subclass—Eulamellibranchia (oysters, clams, shipworms)
Class—Cephalopoda ("head-foot"; cephalopods)
Subclass—Tetrabranchiata (external shells)
Order—Nautiloidea (extinct nautiloids, *Nautilus*)
Order—Ammonoidea (extinct ammonoids)
Subclass—Dibranchiata (internal or absent shells)
Order—Decapoda (squids, cuttlefish, *Sepia*)
Order—Octopoda (octopods)
Order—Belemnoidea (extinct)

PHYLUM—Sipunculida (from Latin, "a little siphon")

Commonly called peanut worms; marine infaunal forms. Sizes from one-quarter inch to 2 feet, bear an introvert proboscis, well-developed systems except circulation, which may be absent. Reproduce sexually, develop from a trochophore larva.

PHYLUM—Annelida (from Latin *annellus,* "ring")

Segmented worms; marine, freshwater, and terrestrial; most forms motile. All organ systems are well developed, except in parasitic forms. Larvae develop as trochophores, or there may be no larval stage.
Class—Polychaeta (marine worms)
Order—Errantia (clamworms, lugworms, sea mouse)

Order—Sedentaria (tube worms, *Chaetopterus*, feather-duster worms)

Class—Oligochaeta (earthworms)

Class—Hirudinea (leeches)

Class—Archiannelida (simple forms)

PHYLUM—Echiuroidea ("adder-tail"; spoon worms)

Unsegmented worms found infaunally in shallow temperate and warm seas. Phylogenetic position uncertain. Larvae may show segmentation. Bear pseudoproboscis, sexes separate, development via trochophore larva.

PHYLUM—Oncopoda ("claw-foot"; miscellany)

A phylum of convenience, usually segmented, claw-bearing animals with unjointed legs. Sexes separate, develop from various larvae. The subphyla may have no real relationship to each other.

SUBPHYLUM—Onycophora (arthropod-annelid link, *Peripatus*)

SUBPHYLUM—Tardigradia ("water-bears")

SUBPHYLUM—Pentastomida (clawed endoparasites)

PHYLUM—Arthropoda ("jointed-legs")

Segmented animals, with jointed legs, usually compound eyes, a chitinous exoskeleton. Abundant in the sea (mostly crustacea), on land (insects and arachnids), and in fresh water (insects and crustacea). Sexes generally separate, many varieties of larval or direct development. Probably the most successful phylum that has ever existed.

SUBPHYLUM—Trilobitomorpha (trilobites and relatives)

SUBPHYLUM—Chelicerata (without antennae, unformed jaws)

Class—Merostomata

Subclass—Eurypterida (eurypterids, extinct)

Subclass—Xiphosuria (horseshoe crabs)

Class—Arachnida (scorpions, spiders, mites, ticks)

Class—Pycnogonida (sea spiders)

SUBPHYLUM—Mandibulata (true jaws and antennae)

Class—Chilopoda (centipedes)

Class—Diplopoda (millipedes)

Class—Insecta (insects, 3/4 million species)

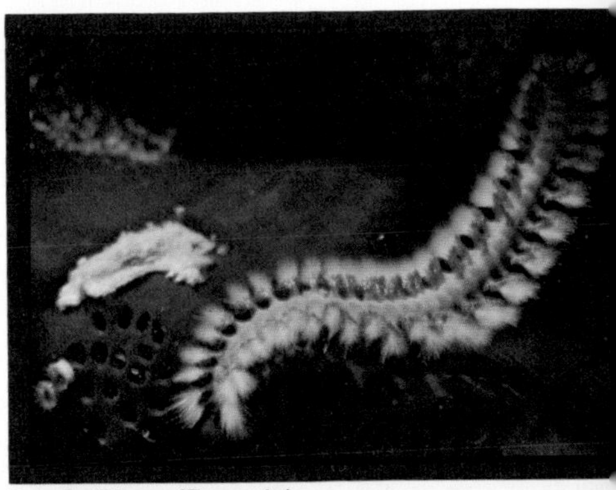

Bristle Worm (Errantia)

Feather-duster Worm (Sedentaria)

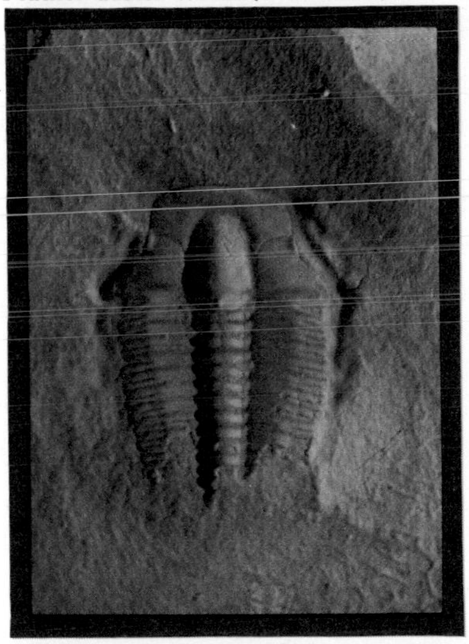

Trilobite (Trilobitomorpha)

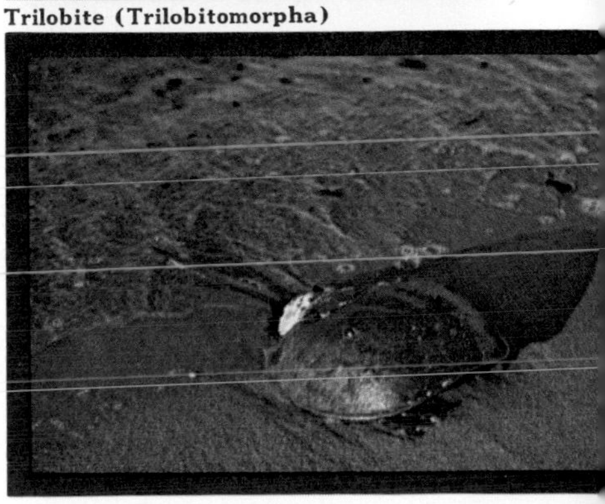

Horseshoe Crab (Xiphosuria)

Crab (Decapoda)

Copepod (Copepoda)

Crinoid (Pelmatozoa)

Starfish and Sea Urchins (Eleutherozoa)

Class—Crustacea (diverse land, sea, and freshwater group)
- *Subclass*—Cephalocarida (primitive forms)
- *Subclass*—Branchiopoda (*Daphnia*, brine shrimp)
- *Subclass*—Ostracoda (bivalve-shelled)
- *Subclass*—Cirripeda (barnacles)
- *Subclass*—Copepoda (copepods)
- *Subclass*—Malacostraca (shrimplike)
  - Superorder—Pancarida (with incompletely fused carapace)
  - Superorder—Peracarida (fused carapace, up to four segments missing)
    - Order—Tanaidacea (tanaids)
    - Order—Isopoda (fish lice, sow bugs)
    - Order—Amphipoda (sand hoppers, beach fleas)
    - Order—Mysidacea (oppossum shrimp)
  - Superorder—Hoplocarida (mantis shrimps)
    - Order—Stomatopoda *(Squilla)*
  - Superorder—Eucarida (with fused carapace)
    - Order—Euphausiacea (krill)
    - Order—Decapoda (shrimp, crabs, lobsters)

PHYLUM—Chaetognatha ("bristle-jaw"; arrow worms)

Marine wormlike forms, possibly related to the nematodes. Common elements of zooplankton, bilaterally symmetrical, lacking vascular, excretory, respiratory organs. Sexual reproduction, individuals hermaphroditic. Epidermis is unique among invertebrates in having more than one layer of cells.

PHYLUM—Pogonophora ("beard-bearer"; beard worms)

Deep-sea forms, unique among higher invertebrates in having no trace of a digestive system—probably a degenerate condition. Considered a relative of the protochordates, they have a very thin body (under one millimeter), up to one foot long. Tube-dwelling, with a three-part body, the anteriormost being tentacles. Unknown before 1933.

PHYLUM—Echinodermata ("spiny-skin")

Exclusively marine, bilateral with possible radial symmetry. Characterized by a water-vascular system, generally advanced organ systems. Development from a bipinnaria larva, or direct development. May be free-living or stalked, rigid or motile.

SUBPHYLUM—Pelmatozoa (stalked forms)
Class—Carpoidea (extinct)
Class—Cystoidea (extinct)
Class—Blastoidea (extinct)
Class—Crinoidea (sea lilies)
Class—Edrioasteroidea (extinct)
SUBPHYLUM—Eleutherozoa (unattached forms)
Class—Holothuroidea (sea cucumbers)
Class—Asteroidea (sea stars, starfishes)
Class—Ophiuroidea (serpent stars, brittle stars)
Class—Echinoidea (sea urchins, sand dollars, heart urchins)

PHYLUM—Chordata (vertebrates, protochordates)

The phylum with the most efficient animals in each of its specialized directions. All have a notochord or embryonic trace. Vertebrates with internal skeletons, advanced organ systems, bilateral symmetry, all have pharyngeal gill slits, dorsal hollow nerve chord; development via various larvae or direct.

SUBPHYLUM—Hemichordata (primitive protochordates)
Class—Pterobranchia (colonial deep-sea forms)
Class—Enteropneusta (acorn worms)
SUBPHYLUM—Cephalochordata (sea lancelets, *Amphioxus*)
SUBPHYLUM—Urochordata (tunicates)
Class—Ascidiacea (sea squirts)
Class—Thaliacea (salps)
Class—Larvacea (tailed neotenous forms)
SUBPHYLUM—Vertebrata (vertebrates)
Class—Agnatha (jawless fish)
Subclass—Cephalaspidomorphi (extinct ostracoderms, single nostril)
Subclass—Pteraspidomorphi (extinct ostracoderms, no nostril)
Subclass—Cyclostomata (lampreys, hagfish)
Order—Myxiniformes (hagfish)
Order—Petromyzontiformes (lampreys)
Class—Placodermi (extinct primitive jawed armored fish)
Order—Athrodira (predatory armored fish)
Order—Antiarchi (small jointed pectoral fins)
Order—Acanthodii ("spiny" sharks)
Order—Stegoselachii ("armored sharks")

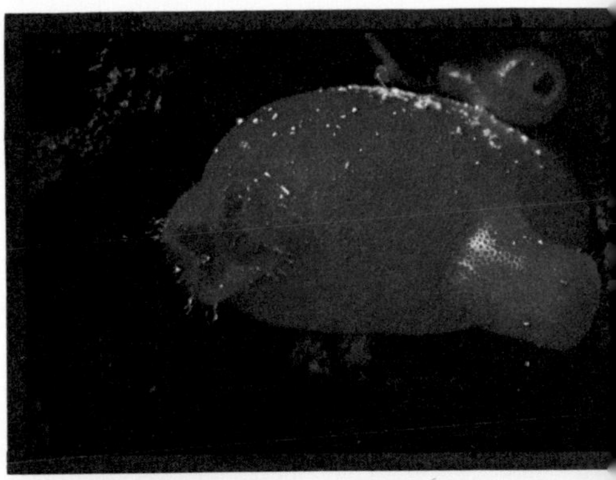
Sea Squirt (Ascidacea)

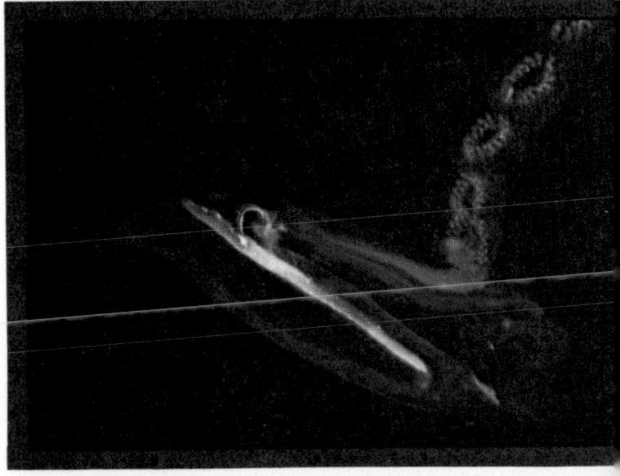
Salps (Thaliacea)

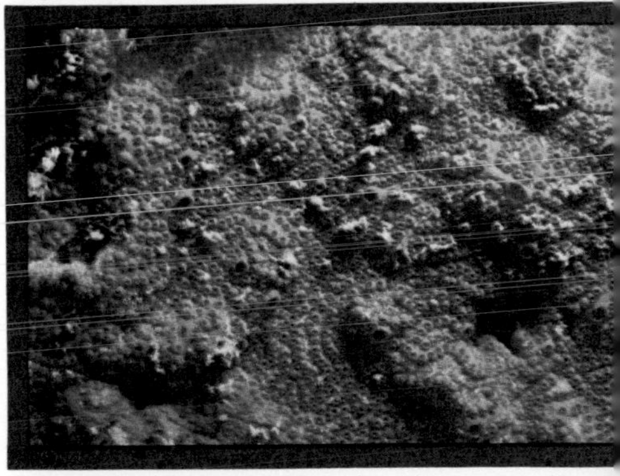
Colonial Ascidians (Ascidacea)

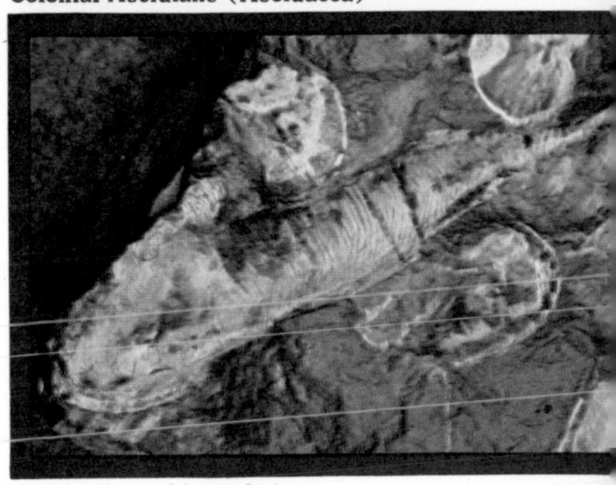
Ostracoderm (Agnatha)

Thresher Shark (Chondrichthyes)

Reef Fish (Osteichthyes)

Regal Anglefish (Osteichthyes)

Parrotfish (Osteichthyes)

- *Class*—Chondrichthyes (cartilaginous fish)
  - *Subclass*—Elasmobranchii (sharks)
    - Order—Squaliformes (sharks, dogfish)
    - Order—Rajiformes (skates, rays)
  - *Subclass*—Holocephali (chimeras)
- *Class*—Osteichthyes (bony fish)
  - *Subclass*—Actinopterygii (ray-finned fish)
    - *Infraclass*—Chondrosteii (primitive forms)
      - Order—Paleonisciformes (extinct)
      - Order—Polypteriformes (bichirs, *Polypterus*)
      - Order—Acipenseriformes (sturgeons, paddlefish)
    - *Infraclass*—Holostei (intermediate forms)
      - Order—Semiontoformes (garpikes and ancestors)
      - Order—Amiiformes (bowfins)
    - *Infraclass*—Teleostei (modern fish)
      - Order—Clupeiformes (herring, anchovies)
      - Order—Elopiformes (tarpons)
      - Order—Anguilliformes (eels)
      - Order—Protacanthopterygii (salmon, trout)
      - Order—Gonorhynchiformes (milkfish)
      - Order—Cypriniformes (minnows, carps, goldfish)
      - Order—Siluriformes (catfishes)
      - Order—Gadiformes (cod, haddock)
      - Order—Batrachoidiformes (toadfish)
      - Order—Lophiiformes (anglerfish)
      - Order—Gobiesociformes (clingfish)
      - Order—Atheriniformes (flying fish)
      - Order—Beryciformes (squirrel fish)
      - Order—Zeiformes (tropical John Dories)
      - Order—Lampridiformes (ribbonfish, oarfish)
      - Order—Gasterosteiformes (sticklebacks, seahorses)
      - Order—Scorpaeniformes (sculpins, sea robins)
      - Order—Pegasiformes (sea moths, sea dragons)
      - Order—Perciformes (tuna, bass, perch, marlin, sunfish, mackerel)
      - Order—Pleuronectiformes (flatfish)
      - Order—Tetraodontiformes (blowfish, trunkfish)
  - *Subclass*—Sarcopterygii (air-breathing fish)
    - Superorder—Crossopterygii (lobe-finned fish)
      - Order—Osteolepiformes (extinct)
      - Order—Coelacanthiformes (recent and extinct coelacanths)

Superorder—Dipnoi (lungfishes)
Class—Amphibia (amphibians)
Subclass—Labyrinthodontia (extinct, *Eryops*)
Subclass—Lepospondyli (extinct)
Subclass—Gymnophiona (wormlike forms)
Subclass—Urodela (most living amphibians)
Order—Proteida (permanent-gilled, with lungs)
Order—Meantes (permanent-gilled, no lungs)
Order—Mutabilia (adults without gills)
Subclass—Anuria (frogs, toads)
Class—Reptilia (reptiles)
Subclass—Anapsida (solid skull roof)
Order—Chelonia (turtles)
Order—Cotylosauria (stem reptiles, extinct)
Order—Mesosauria (extinct aquatic reptiles)
Subclass—Euryapsida (marine reptiles)
Order—Protosauria (extinct)
Order—Sauropterygia (extinct, plesiosaurs, mosasaurs)
Order—Ichthyosauria (ichthyosaurs)
Subclass—Diapsida (ruling reptiles)
Infraclass—Lepidosauria (snakes, lizards, and their ancestors)
Infraclass—Archosauria (crocodilians, dinosaurs)
Order—Thecodontia (extinct crocodiles)
Order—Crocodilia (crocodiles, alligators, caymens)
Order—Saurischia (*Tyrannosaurus, Brontosaurus*)
Order—Ornithiscia (duck-billed, armored dinosaurs)
Order—Pterosauria (flying reptiles)
Class—Aves (birds)
Subclass—Archaeornithes (extinct, *Archaeopteryx*)
Subclass—Neornithes (true birds)
Superorder—Odontognathae (extinct toothed forms)
Superorder—Ichthyornithes (extinct billed birds)
Superorder—Paleognathae (flightless birds, ostriches, moas, kiwis)
Superorder—Neognathae (flying birds)
Order—Sphenisciformes (penguins)
Order—Procellariformes (petrels, albatrosses)
Order—Gaviiformes (loons, diving birds)
Order—Podiciformes (grebes)

Jacks (Osteichthyes)

Turtle (Chelonia)

Iguana (Lepidosauria)

Penguins (Sphenisciformes)

Cormorants (Pelecaniformes)

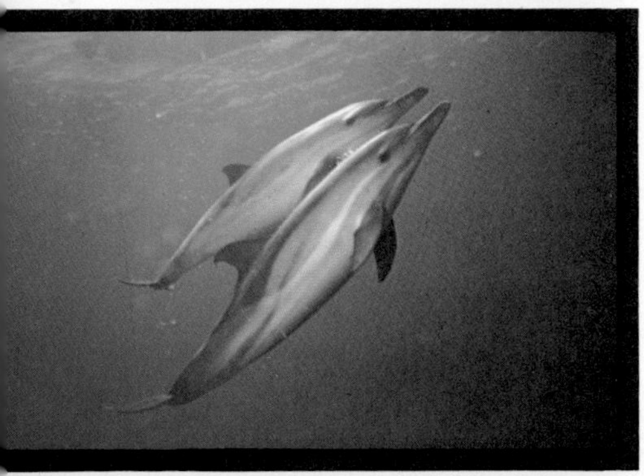

Dolphin (Odontoceti)

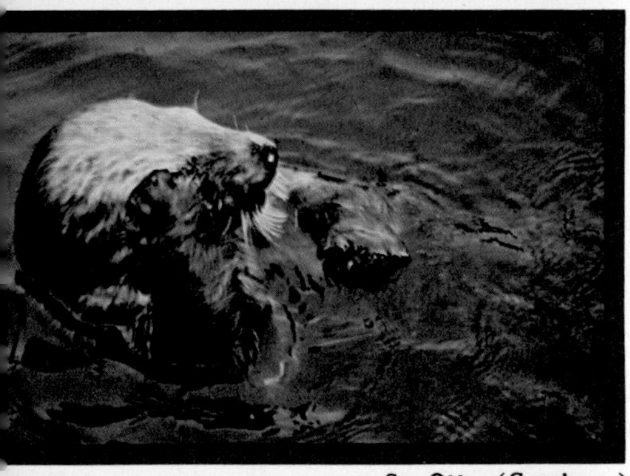

Sea Otter (Carnivora)

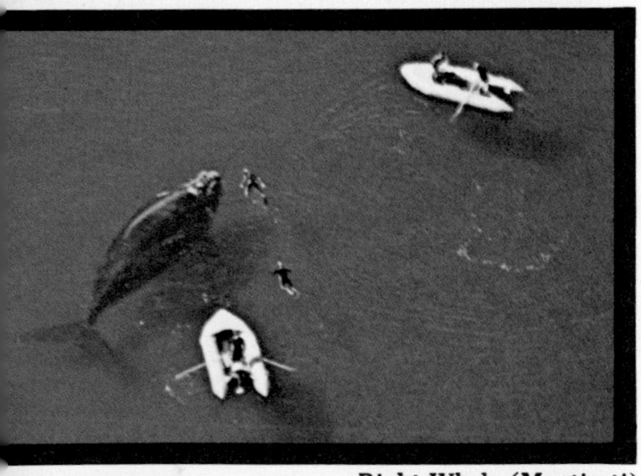

Right Whale (Mysticeti)

Order—Pelecaniformes (gannets, cormorants, pelicans, frigate birds)
Order—Ciconiformes (storks, herons, ibises, flamingos)
Order—Anseriformes (swans, ducks, geese)
Order—Falconiformes (eagles, hawks, vultures, kites, osprey)
Order—Galliformes (chickens, turkeys, quail, grouse)
Order—Gruiformes (cranes, coots)
Order—Charadriiformes (gulls, terns, plover)
Order—Psitticaformes (parrots)
Order—Columbiformes (pigeons)
Order—Cuculiformes (cuckoos)
Order—Strigiformes (owls)
Order—Caprimulgiformes (whippoorwills)
Order—Apodiformes (hummingbirds)
Order—Coraciiformes (kingfishers)
Order—Piciformes (woodpeckers)
Order—Passeriformes (thrushs, crows, larks, jays, most song birds)

Class—Mammalia (mammals)
- *Subclass*—Eotheria (extinct mesozoic ancestors)
- *Subclass*—Prototheria (egg layers, platypus, echidna)
- *Subclass*—Allotheria (extinct cenozoic mammals)
- *Subclass*—Theria (most modern mammals)
  - Superorder—Marsupialia (pouched animals)
  - Superorder—Eutheria (placental mammals)
    - Order—Insectivora (insectivores, shrews, moles, hedgehogs)
    - Order—Chiroptera (bats)
    - Order—Primates (monkeys, apes, man)
      - Suborder—Anthropoidea (anthropoid apes, man)
        - Family—Hominidae (man)
    - Order—Edentata (sloths, armadillos, anteaters)
    - Order—Rodentia (rodents, rats, mice, porcupines)
    - Order—Lagomorpha (rabbits, hares, pikas)
    - Order—Cetacea (porpoises and whales)
      - Suborder—Archaeoceti (ancestral whales, *Zeuglodon*)
      - Suborder—Odontoceti (toothed whales, dolphins, orcas)
      - Suborder—Mysticeti (rorquals)

Order—Carnivora (felines, canines, seals)
    Suborder—Fissipedia (four-footed carnivores)
    Suborder—Pinnipeda (marine carnivores)
        Family—Otariidae (fur seals)
        Family—Odebenidae (walruses)
        Family—Phocidae (hair seals)
Order—Tubilidentata (aardvarks)
Order—Hyracoidea (coneys)
Order—Proboscoidea (elephants)
Order—Sirenia (sea cows, manatees, dugongs)
Order—Perissodactyla (odd-toed ungulates, horses, rhinoceros)
Order—Artiodactyla (even-toed ungulates, cattle, pigs, camels)

Seal (Pinnipeda)

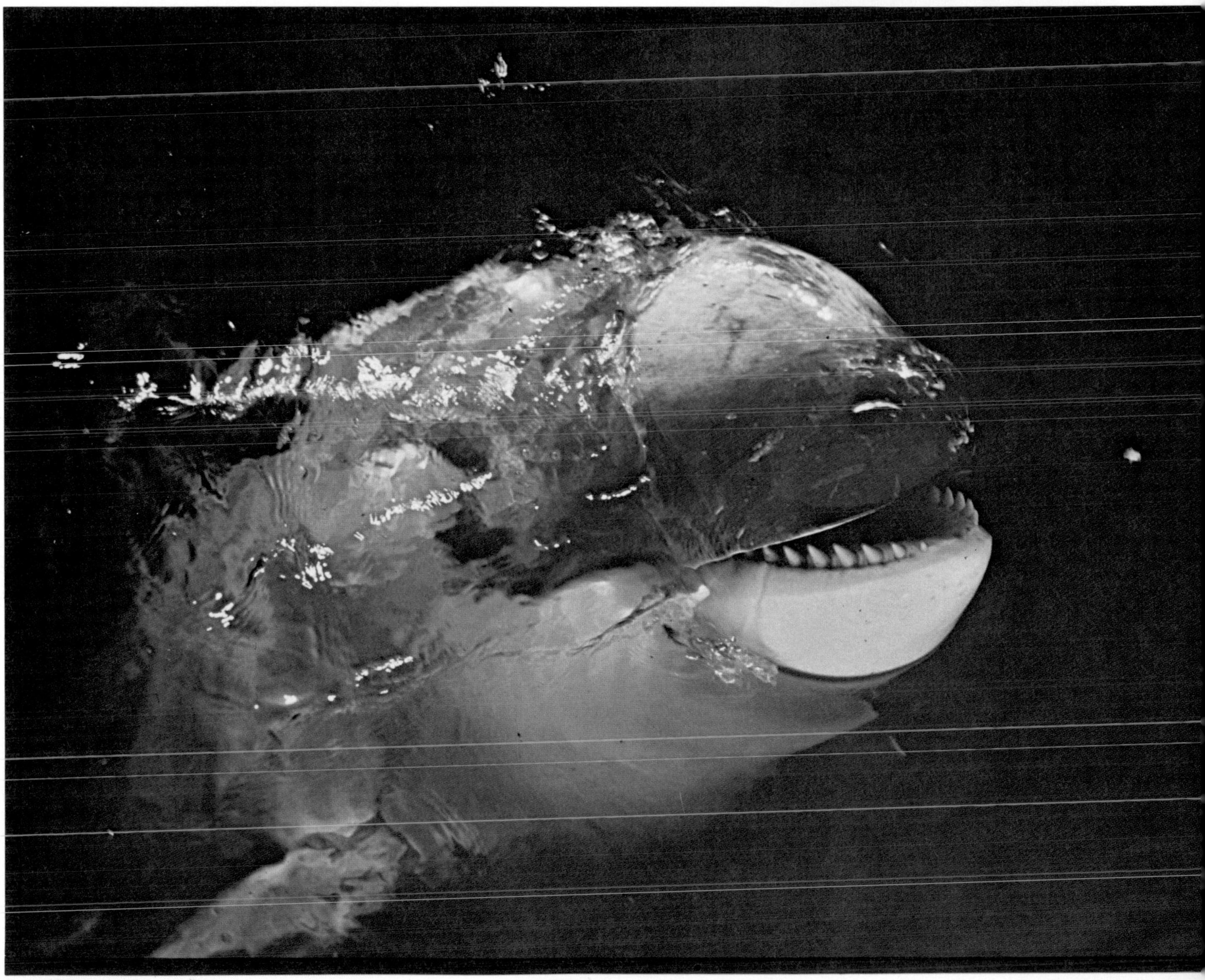

Orca (Odontoceti)

Chapter VII. Time and Measure

From a snail's view, a mile is a life's journey; to the albatross the entire earth is a season's jaunt. Astronauts orbit the planet in one hour and 40 minutes. In essence we impose a time and measure system on all that exists in the known world, but we forget that it is oriented around the perception of an approximately 5-foot, 100-pound human who may have lived an average of 50 years.

In this chapter simple information is gathered about the common systems of measures used by marine scientists, and some useful figures may help to put our earth in perspective for the student of the oceans as well as for all those who are interested in the sea.

The Metric System

The system of weights and measures used in England, the United States, and territories that are or were under their influence is generally called the British system. It was the earliest rigidly standardized European system and it helped England become the major industrial and scientific force of the eighteenth and nineteenth centuries.

Even though it was widely used the British system had faults in common with the earlier Greek, Roman, Egyptian, and Assyrian systems. It was based on random units, which had uneven relationships to each other. The foot was the length of precisely that—the king's foot. The inch was the length of three barleycorns laid end to end. The yard was the length of the king's arm. Short kings made for short measures. The temperature scale was based on a more reproducible unit—the freezing and boiling points of pure water. But still the scales of temperature in the British system have little in common with the system of weights, and one suspects that Daniel Fahrenheit (1686–1736) devised the scale to best fit in with the first thermometer he made.

The British weight units seem to trace back to very ancient times. The pound is a prehistoric concept, which was adapted by the Romans, Goths, and then by the English in the early Middle Ages. The ounce was originally one-twelfth of a pound (and still is in troy, or apothecaries', weights), and finally was standardized as the familiar one-sixteenth of a pound.

A quick survey will show that intervals for measures in the British system include fractions of one-third, one-twelfth, one-sixteenth, and some truly strange numbers like 5280 feet to the mile. There are also some rather arcane units in common usage in the British system. Many of them have obvious origins—such as the rod, chain, stone (which may vary from 4 to 26 pounds), and hand (as a unit in measuring horses). Others are more obscure—acres, bushels, pecks, perches, hogsheads, barrels, pints, quarts, carats, and grosses. For better communication among scientists, cartographers, and tradesmen, a new system had to be found.

In the late eighteenth century in France, the Académie Royale des Sciences proposed that a new system should be developed; the academy tested and finally produced the metric system (after a minor interruption by the French Revolution). The basic unit was the meter, a distance one ten-millionth part of the quarter of a great circle meridian of the earth. The greatest standardization in the metric system is in the division of units

*A diver **engulfed in fish,** off the tip of Baja, California, can get a feel for the magnitude of life in the sea and the difficulty in measuring its abundance.*

—all based on decimals—with a few prefixes, added to the long list of suffixes, any quantity can be simply indicated.

The weight measure and volume units are related to each other by a simple concept. The milliliter is equal in quantity to the amount of water in one cubic centimeter. This also equals one gram, the basis for the weight system. The water must be at its maximum density—about 4° centigrade. The centigrade scale is related to the Fahrenheit scale; however, the boiling and freezing points of pure water are rather logically placed at the 0° and 100° marks.

Today almost all scientists communicate data using the metric system, as do people in most of the world. The meter has been standardized as two lines etched on a platinum-iridium bar, shown to equal 1,650,-763.73 wavelengths of the orange-red light given off by krypton^{86} in an excited state. For those who hang on to older systems, the cost of the converting tools may seem paramount at the moment, but eventually they too must march to the metric drum.

*The **Telenaute,** an unmanned submersible, explores the mysterious Fontaine de Vaucluse.*

Metric Conversion Table

Metric Prefixes

nano—one-billionth	*hecto*—one hundred
micro—one-millionth	*kilo*—one thousand
centi—one-hundredth	*mega*—one million
deci—one-tenth	*giga*—one billion
deca—ten	

Weights

1 pound =	.4536 kilograms
1 kilogram =	2.205 pounds

(approximately 2 pounds are 1 kilogram less 10%)

1 ounce =	31.10 grams
1 gram =	0.035 ounces
1 ton =	.907 metric tons
1 metric ton =	1.102 tons

Length

1 foot =	.305 meters
1 meter =	3.28 feet

(approximately 3 feet plus 10% to each meter)

1 inch =	2.54 centimeters
1 centimeter =	.394 inches
1 centimeter =	.033 feet
1 yard =	.848 meters
1 mile =	1.61 kilometers
1 kilometer =	.621 miles
1 fathom =	1.83 meters

Capacity

1 quart =	.946 liters (liquid) or 1.101 liters (dry)
1 liter =	1.057 quarts (liquid) or .908 quarts dry
1 pint =	.473 liters (liquid) or .551 liters (dry)
1 deciliter =	.21 pints (liquid) or .18 pints (dry)
1 decaliter =	2.64 gallons (liquid) or 1.14 pecks (dry)
1 bushel =	35.238 liters
1 hectoliter =	26.4 gallons (liquid) or 2.84 bushels (dry)
1 centiliter =	.338 fluidounces or .6 cubic inches
1 milliliter =	.27 fluidrams or .06 cubic inches

Volume

1 cubic foot =	.0283 cubic meters
1 cubic meter =	35.31 cubic feet
1 cubic inch =	16.39 cubic centimeters
1 cubic centimeter =	.061 cubic inches
1 cubic yard =	.7646 cubic meters
1 cubic meter =	1.308 cubic yards
1 cubic mile =	4.17 cubic kilometers
1 cubic kilometer =	.2388 cubic miles
1 stere =	1.31 cubic yards

Temperature

To convert Fahrenheit to centigrade (Celsius) temperature, subtract 32°, then multiply by 5/9 (or divide by 2 and add 10%).

To convert centigrade to Fahrenheit, multiply by 9/5 (or double and subtract 10%), then add 32°. (e.g., 212° F. minus 32° = 180° F.; multiply by 5/9 = 100° C.)

Area

1 square foot =	0.93 square meters
1 square meter =	10.76 square feet
1 square mile =	2.59 square kilometers
1 square kilometer =	.386 square feet
1 acre =	.405 hectares
1 hectare =	2.47 acres

Miscellaneous

1 knot =	1.15 miles/hour or 1852 kilometers/hour
1 radian =	57.29 degrees of a circle (2 pi radians equal 360°)
1 circle =	400 grads divided into decigrads, centigrads, etc.
1 fathom =	six feet
1 micron =	one one-millionth of a meter
1 angstrom =	one-ten-billionth of a meter

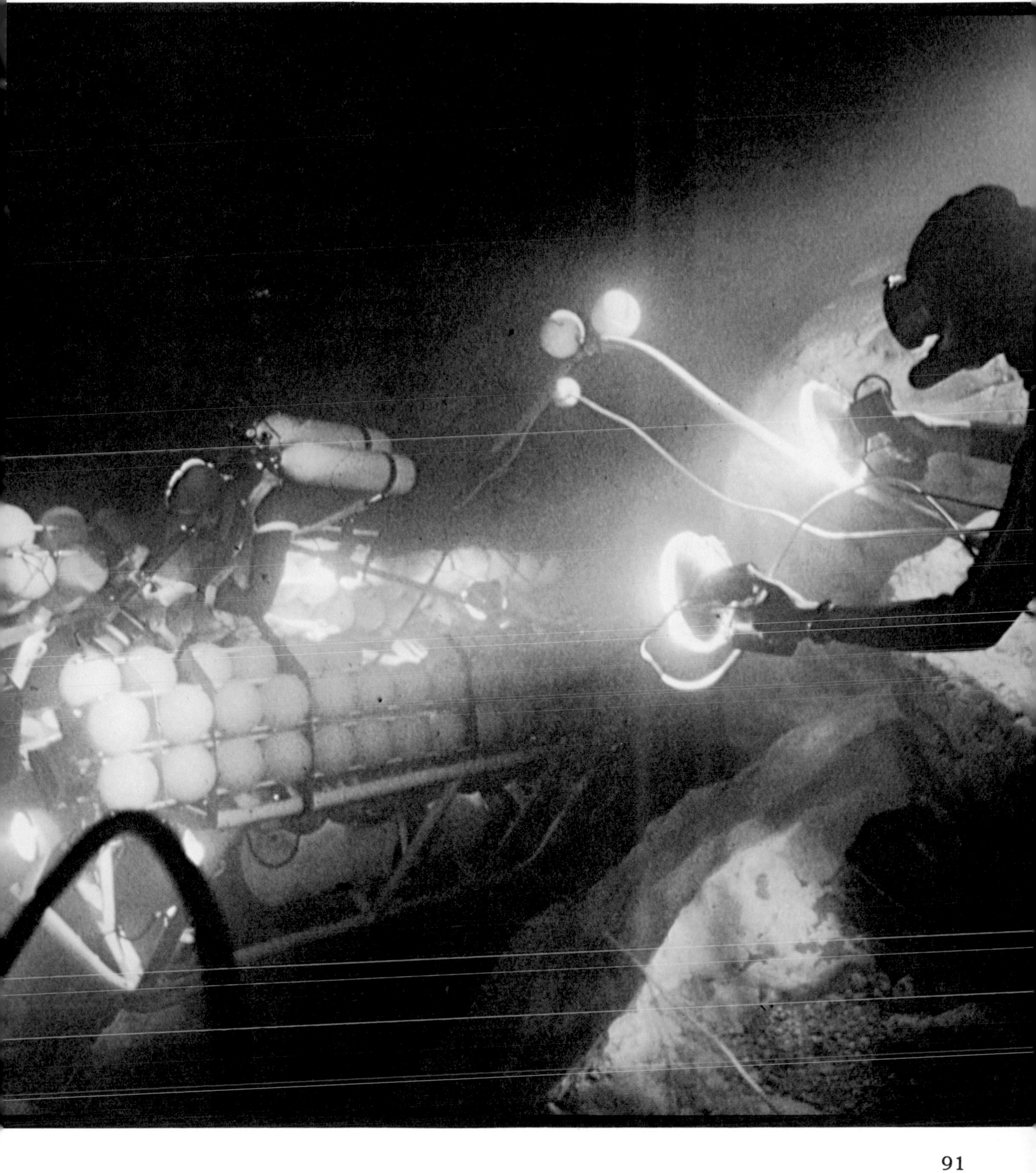

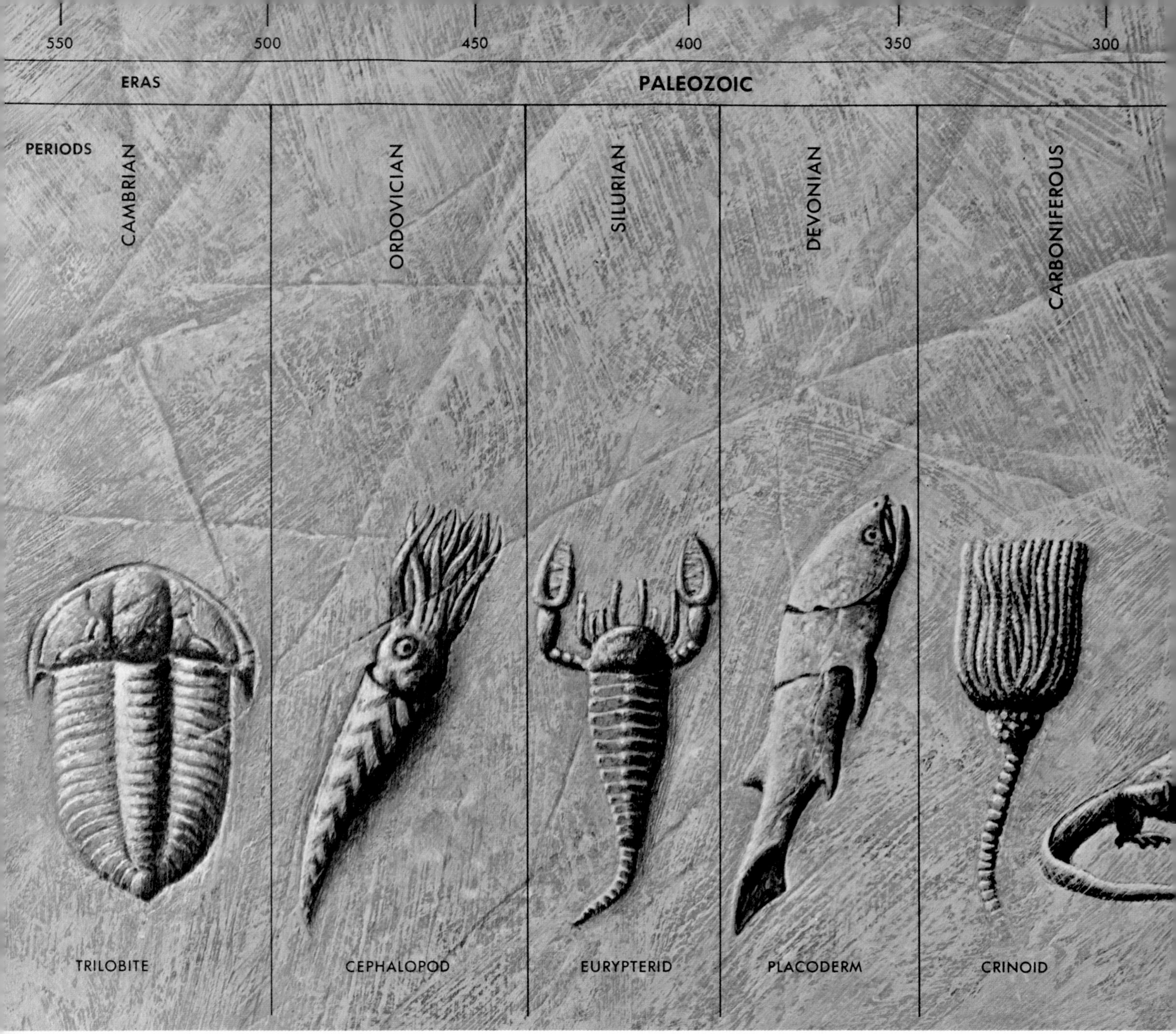

Earth Time

The span of recorded evolution on earth—that is, the time interval bracketed by fossil evidence—has been divided into convenient units by geologists. The largest of these divisions are *eras,* whose boundaries correspond to major catastrophic or evolutionary events. Within eras are *periods,* bounded by natural or lithologic events on a smaller scale. Sometimes a period boundary (and occasionally an era boundary) is not determinable in a particular location; for example, the Permian–Triassic period boundary and the break between the Paleozoic and Mesozoic eras cannot be accurately documented in the terrestrial beds of the southwestern United States. However, the marine animals of this time underwent major upheavals, and corresponding marine deposits clearly show the break in fossils.

The most recent era, the Cenozoic, can be handled several ways, depending on personal interpretations. If we are still in the

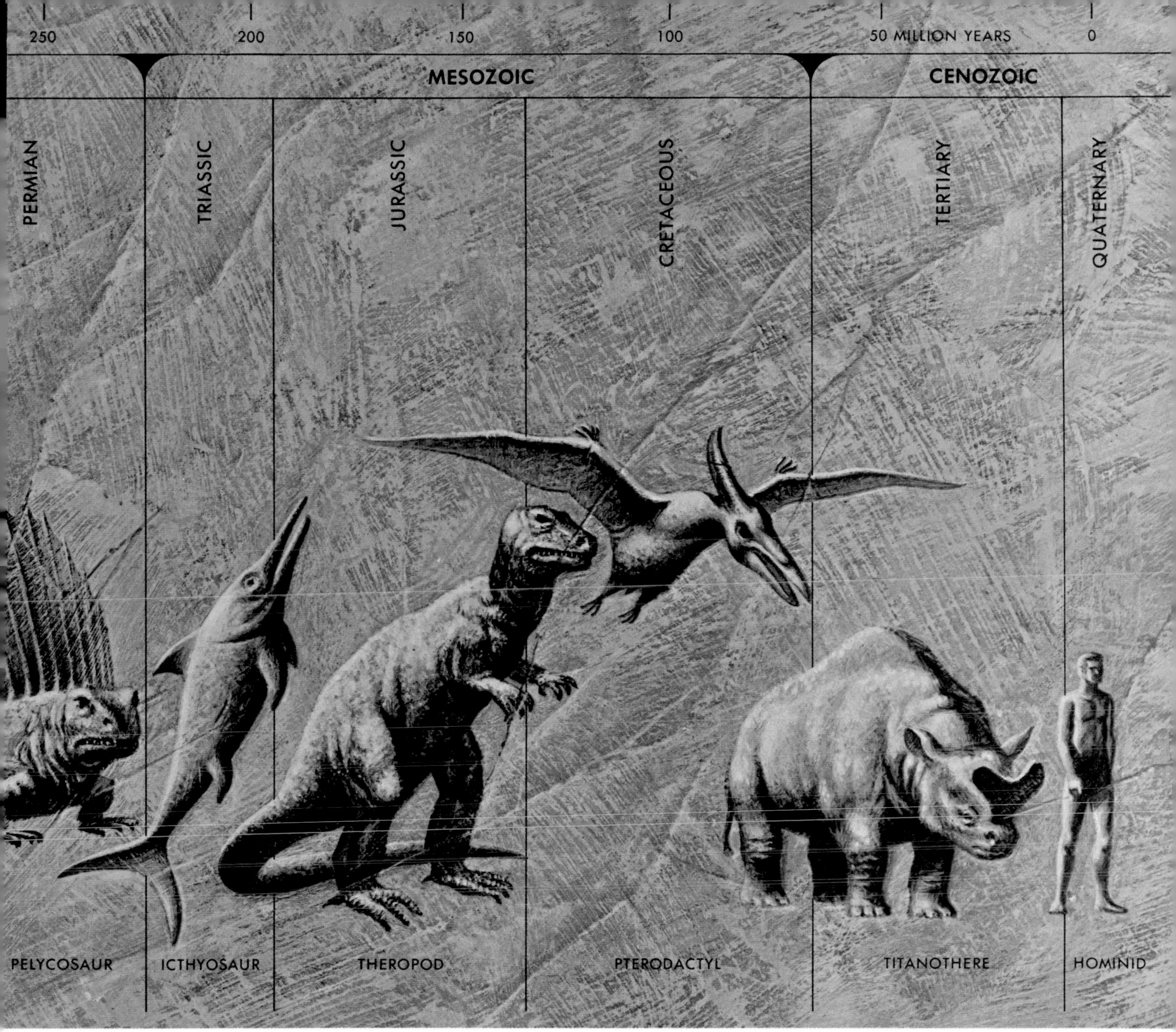

Ice Age, then we are in the Pleistocene epoch. *Epochs* are a subdivision used only for the Cenozoic era. However, if we are out of the Ice Age and in a new and unique time, then we are in the Recent period and may even be in a new era. The chart above shows the most generally agreed upon geologic time scale, with two periods in the Cenozoic—the Tertiary and Quaternary. The Quarternary period includes the Pleistocene epoch and the Recent (or what is sometimes called the "Holocene") epoch.

The Geologic Time Scale *(above) divides the immense amounts of time that life on earth has been in existence into parcels we can deal with.*

The earth existed at least four billion years before the first animal started to leave good fossils. For convenience, that entire span of time—over six times as long as the time indicated above—is generally referred to as the Precambrian. New evidence of life during this time is being gathered right now, and soon a meaningful breakdown for the Precambrian era will emerge.

The Numbers of Things

The geologic time scale that we have illustrated includes only the portion of the earth's history for which we have good traces of life. The portion of earth history before the Cambrian period is logically called the Precambrian. The earth is thought to be 4.6 billion years old; this number is almost impossible to appreciate, even though the word *billion* is part of everyday speech. A billion years is 50 million human generations; 500 times the span of the human race, 100,000,000,000,000 times the life span of some single-celled organisms. For convenience' sake, scientists put large numbers such as the above into exponents of 10 (for example, the above would be 10^{14}; the number 50 is 5 times 10^{1}, .74 is 7.4 times 10^{-1}). They deal with numbers as large as a googol, which is 10^{100}, and one suspects that the strange name was given because it is impossible to get a feel for a number that size.

There are many bits of data which are of interest concerning the earth and sea; one should keep in mind that it is easy to rattle off numbers which are so large that they are virtually inconceivable. The data presented here is given as reference, not as an attempt to impress the superficial reader. Nevertheless, some of it is fantastic.

The Seas

Age—The oceans cannot be precisely dated, but there are marine sediments at least 3.5 billion years old, and the seas obviously cannot be older than the earth itself.

Surface Area—(70.78% of the earth's surface) 139.4 × 10^6 sq mi (361 × 10^6 sq km).

Surface Area over Continental Shelves—10.594 × 10^6 sq mi (27.438 × 10^6 sq km). This is approximately 7.6% of the total sea surface and 5.38% of the earth's total surface; an area comparable to South America and Europe combined.

Surface Area over Continental Slopes—21.328 × 10^6 sq mi (55.243 × 10^6 sq km), accounting for approximately 15.3% of the sea surface and 10.83% of the earth's total surface; an area almost twice the land surface of Africa.

Surface Area over Abyssal Plains—105.805 × 10^6 sq mi (274.044 sq km). This is approximately 75.9% of the sea's total surface and 53.72% of the earth's total surface; more than 6 times the surface area of the Asian continent.

Surface Area over Trenches—(depths below 6000 m) 1.673 × 10^6 sq mi (4.334 × 10^6 sq m). This is approximately 1.2% of the total sea surface and 0.85% of the earth's surface; slightly more than half the surface area of Australia.

Surface Area of Oceans and Seas—*Pacific Ocean:* 63.8 × 10^6 sq mi (165.242 × 10^6 sq km); *Atlantic Ocean:* 31.83 × 10^6 sq mi (82.44 × 10^6 sq km); *Indian Ocean:* 28.36 × 10^6 sq mi (73.45 × 10^6 sq km); *Arctic Ocean:* 5.44 × 10^6 sq mi (14.09 × 10^6 sq km); *Caribbean Sea:* 1.063 × 10^6 sq mi (2.754 × 10^6 sq km); *Mediterranean Sea:* 9.667 × 10^5 sq mi (2.504 × 10^6 sq km); *Bering Sea:* 8.757 × 10^5 sq mi (2.268 × 10^6 sq km); *Gulf of Mexico:* 5.957 × 10^5 sq mi (1.543 × 10^6 sq km); *Sea of Okhotsk:* 5.898 × 10^5 sq mi (1.528 × 10^6 sq km); *East China Sea:* 4.823 × 10^5 sq mi (1.249 × 10^6 sq km); *Hudson Bay:* 4.758 × 10^5 sq mi (1.232 × 10^6 sq km); *Sea of Japan:* 3.891 × 10^5 sq mi (1.008 × 10^6 sq km); *Andaman Sea:* 3.079 × 10^5 sq mi (7.976 × 10^5 sq km); *North Sea:* 2.221 × 10^5 sq mi (5.753 × 10^5 sq km); *Black Sea:* 1.784 × 10^5 sq mi (4.620 × 10^5 sq km); *Red Sea:* 1.691 × 10^5 sq mi (4.379 × 10^5 sq km); *Baltic Sea:* 1.631 × 10^5 sq mi (4.223 × 10^5 sq km).

Volume of World Ocean—330 × 10^6 cu mi (1370.323 × 10^6 cu km).

Volume of the Oceans—*Pacific Ocean:* 173.688 × 10^6 cu mi (723.699 cu km); *Atlantic Ocean:* 81.047 × 10^6 cu mi (337.699 cu km); *Indian Ocean:* 70.066 × 10^6 cu mi (291.945 × 10^6 cu km); *Arctic Ocean:* (including Canadian Archipelago, Baffin Bay, and the Norwegian Sea) 4.075 × 10^6 cu mi (16.980 cu km).

Weight of the Water—0.0015 × 10^{21} tons (.0013 × 10^{21} metric tons).

Mean Salinity of the Ocean—35 parts per thousand ($^0/_{00}$).

Highest Salinity—in the Red Sea, 40 $^0/_{00}$.

Lowest Salinity—Gulf of Bothnia, 5 $^0/_{00}$.

Salt Content of the Sea—1 cu mi of seawater contains 1.66 × 10^8 tons (1.506 × 10^8 metric tons) of salt; total is enough to cover the continents with a layer 500 ft (152.5 m).

Elements of the Ocean

(per 1 cu mi [4.167 cu km] of seawater)

Element	English System	Metric System
oxygen	4.037×10^9 tons	3.662×10^9 metric tons
hydrogen	5.09×10^8 tons	3.662×10^8 metric tons
chlorine	89.5×10^6 tons	81.178×10^6 metric tons
sodium	49.5×10^6 tons	44.897×10^6 metric tons
magnesium	6.125×10^6 tons	5.555×10^6 metric tons
sulfur	4.24×10^6 tons	3.846×10^6 metric tons
calcium	1.88×10^6 tons	1.705×10^6 metric tons
potassium	1.79×10^6 tons	1.624×10^6 metric tons
bromine	3.06×10^5 tons	2.775×10^5 metric tons
carbon	1.32×10^5 tons	1.197×10^5 metric tons
strontium	3.77×10^4 tons	3.419×10^4 metric tons
boron	2.26×10^4 tons	2.05×10^4 metric tons
silicon	1.413×10^4 tons	1.282×10^4 metric tons
fluorine	6.125×10^3 tons	5.555×10^3 metric tons
argon	2.825×10^3 tons	2.562×10^3 metric tons
nitrogen	2.35×10^3 tons	2.131×10^3 metric tons
lithium	940 tons	852.6 metric tons
rubidium	565 tons	512.5 metric tons
phosphorus	330 tons	299.3 metric tons
iodine	235 tons	213.1 metric tons
indium	94 tons	85.3 metric tons
aluminum	47 tons	42.6 metric tons
iron	47 tons	42.6 metric tons
molybdenum	47 tons	42.6 metric tons
zinc	47 tons	42.6 metric tons
barium	29 tons	26.3 metric tons
arsenic	14 tons	12.7 metric tons
copper	14 tons	12.7 metric tons
lead	14 tons	12.7 metric tons
protactinium	14 tons	12.7 metric tons
selenium	14 tons	12.7 metric tons
tin	14 tons	12.7 metric tons
vanadium	9.4 tons	8.53 metric tons
manganese	9.4 tons	8.53 metric tons
titanium	4.7 tons	4.26 metric tons
thorium	3.3 tons	3.49 metric tons
cesium	2.4 tons	2.18 metric tons
antimony	2.4 tons	2.18 metric tons
cobalt	2.3 tons	2.09 metric tons
nickel	2.3 tons	2.09 metric tons
cerium	1.8 tons	1.63 metric tons
yttrium	1.4 tons	1.27 metric tons
silver	1.4 tons	1.27 metric tons
lanthanum	1.4 tons	1.27 metric tons
krypton	1.4 tons	1.27 metric tons
neon	1.4 tons	1.27 metric tons
bismuth	1.885×10^3 lb	855.79 kg
tungsten	940 lb	426.76 kg
xenon	940 lb	426.76 kg
germanium	565 lb	256.51 kg
cadmium	518 lb	235.17 kg
chromium	470 lb	213.38 kg
scandium	377 lb	171.16 kg
mercury	280 lb	127.12 kg
gallium	280 lb	127.12 kg
tellurium	94 lb	42.68 kg
niobium	47 lb	21.34 kg
helium	47 lb	21.34 kg
gold	38 lb	17.25 kg
radium	3×10^{-4} lb	1.36×10^{-4} kg
radon	9×10^{-8} lb	4.09×10^{-8} kg

Mean Depth of the Sea—12,464 ft (3800 m).

Deepest Area—*Pacific Ocean:* Philippine Trench, 37,782 ft (11,523.51 m); *Atlantic Ocean:* Puerto Rico Trench, 27,498 ft (8386.89 m); *Indian Ocean:* Diamantina Depth, 26,400 ft (8,052 m); *Arctic Ocean:* area unnamed, 17,880 ft (5453.4 m); *Bering Sea:* Aleutian Trench, 25,194 ft (7684 m); *Caribbean Sea:* Cayman Trench, 23,288 ft (7100 m); *Mediterranean Sea:* Hellenic Trough, 16,702 ft (5092 m); *Gulf of Mexico:* area unnamed, 14,358 ft (4379 m); *Andaman Sea:* area unnamed, 13,700 ft (4179 m); *Sea of Japan:* area unnamed, 13,281 ft (4049 m); *Sea of Okhotsh:* area unnamed, 11,067 ft (3374 m); *East China Sea:* Okinawa Trough, 8912 ft (2717 m); *Red Sea:* area unnamed, 7740 ft (2361 m); *Black Sea:* area unnamed, 7364 ft (2245 m); *North Sea:* area unnamed, 2052 ft (626 m); *Baltic Sea:* area unnamed, 1506 ft (459 m); *Hudson Bay:* area unnamed, 846 ft (258 m).

Greatest Tides—Bay of Fundy, Nova Scotia, 53 ft (16.2 m).

Strongest Currents—The Gulf Stream, traveling north along the east coast of the United States, and the Kuroshio Current, which moves north along the east coast of the Philippines, Formosa, and Japan, each have surface velocities over 80 in per sec (200 cm per sec).

Weakest Currents—Abyssal currents in the Pacific Ocean have been estimated to move hundredths of a cm per sec.

Highest Waves—Winds of hurricane force have produced waves from 75–90 ft high (23–28 m) in the North Atlantic. Storm waves in the Pacific have been measured at over 100 ft (31 m). Tsunamis reach nearly 200 ft (61 m) high.

Amount and Percent of Total Water—*ice:* 7×10^6 cu mi (2.91×10^7 cu km), 2.15%; *fresh water:* 2.046×10^6 cu mi (8.51×10^6 cu km), 0.63%; *salt water:* 3.2×10^8 cu mi (1.33×10^9 cu km), 97.2%; *vapor:* 3.1×10^3 cu mi (1.29×10^4 cu km), 0.001%.

Density of Ice—At 32° F. (0° C.) ice is about 9% less dense than water; therefore, ice floats in water. Ice becomes less dense as temperature increases.

The Earth

Diameter—7926 mi (12,761 km) at the equator, 7900 mi (12,719 km) at the poles.

Circumference—approximately 24,800 mi (39,928 km) at the poles; slightly larger at the equator.

Surface Area, Total—197×10^6 sq mi (510.23×10^6 sq km).

Land Area—(29.22% of total) 57.6 × 10^6 sq mi (149.184 × 10^6 sq km).

Weight—6.595 × 10^{21} tons (5.982 × 10^{21} metric tons). The core weighs 2.08 × 10^{21} tons (1.88 × 10^{21} metric tons); the mantle 4.49 × 10^{21} tons (4.07 × 10^{21} metric tons); and the crust, including land and sea, .029 × 10^{21} tons (.026 × 10^{21} metric tons). There are about .00025 × 10^{21} tons (.00023 × 10^{21} metric tons) of glacier ice on earth.

Highest Point—Mount Everest, Nepal–Tibet, 29,028 ft (8853.5 m).

Lowest Point—Dead Sea, Israel–Jordan, −1286 ft (−392.2 m).

Largest Island—Greenland, 8.4 × 10^5 sq mi (2.18 × 10^6 sq km); almost equal in size to the North American continent.

Longest River—Nile, Africa, 4145 mi (6673.5 km); slightly less than the distance from New York City to Naples, Italy.

Largest Lake—Caspian Sea, Iran–USSR, 143,550 sq mi (371,795 sq km); an area comparable to Great Britain, Ireland, Belgium, and the Netherlands.

Wettest Spot—Mount Waialeale, Hawaii, averaging 471 in (1196.3 cm) of precipitation per year. Cherrapunji, India, holds the one-year record of 1042 in (2646.7 cm), set in 1861.

Driest Spot—Atacama Desert, Chile, rainfall barely measurable. At nearby Calama, Chile, no rain has ever been recorded.

Coldest Spot—Vostok, Antarctica, −127° F. (−88.4° C.), recorded in 1960.

Hottest Spot—Al'Aziziyah, Libya, 136° F. (57.8° C.), recorded in 1922.

Strongest Surface Wind—231 mph (371.9 kmph), recorded on Mount Washington, New Hampshire, in 1934.

The Sun

Volume—1,304,000 times the volume of the earth.

Mass—332,950 times the mass of the earth.

Diameter—860,017 mi (1,384,627 km).

Mean Distance from Earth—92,960,197.2 mi (149,665,917.5 km).

Life on Earth

Oldest Known Organisms—blue-green algae, 3.4 billion years old.

Oldest Known Animals—questionable fossil jellyfish, 1 billion years old.

Largest Animal That Ever Lived—blue whale.

Longest Organism—giant kelp, *Macrocystis.*

Longest-Lived Species—brachiopod, *Lingula*, which appeared during the Cambrian period, 550,000,000 years ago.

Number of Animal Species—1.5 million (over half are insects); there may be as many as 10 million undescribed (worms, insects, etc.).

Number of Individual Organisms on Earth—perhaps a googol (10^{100}).

Amount of Animal Life Removed from the Sea—(data from 1971) *herring, sardines, anchovies,*

etc.: 2.09×10^7 tons (1.896×10^7 metric tons); *cod, hake, haddock,* etc.: 1.18×10^7 tons (1.073×10^7 metric tons); *flounder, halibut, sole,* etc.: 1.50×10^6 tons (1.36×10^6 metric tons); *salmon, trout, smelts,* etc.: 2.38×10^6 tons (2.16×10^6 metric tons); *tuna, bonito, billfish,* etc.: 1.76×10^6 tons (1.6×10^6 metric tons); *tuna, bonito, billfish,* etc.: 1.76×10^6 tons (1.6×10^6 metric tons); *molluscs:* 3.54×10^6 tons (3.21×10^6 metric tons); *crustaceans:* 1.85×10^6 tons (1.68×10^6 metric tons).

Diving Capabilities of Some Animals: (In most instances figures represent actual measurements; animals may very well travel deeper.) *Sperm whale:* over 1 mi (1.6 km); *blue whale:* over 1000 ft (300 m); *porpoise:* 164 ft (50 m); *bottle-nosed dolphin:* over 650 ft (200 m); *fin whale:* 1640 ft (500 m); *walrus:* 295 ft (90 m); *Weddell seal:* 1968 ft (600 m); *northern fur seal:* 230 ft (70 m); *harp seal:* 902 ft (275 m); *northern elephant seal:* 600 ft (183 m); *gray seal:* 475 ft (145 m); *man:* 260 ft (79 m).

Number of Atolls—Approximately 400 fully developed atolls scattered throughout tropical oceans.

Both ***Calypso and icebergs*** *float because they displace a weight of water greater than their own.*

▲A

▲B

▼C

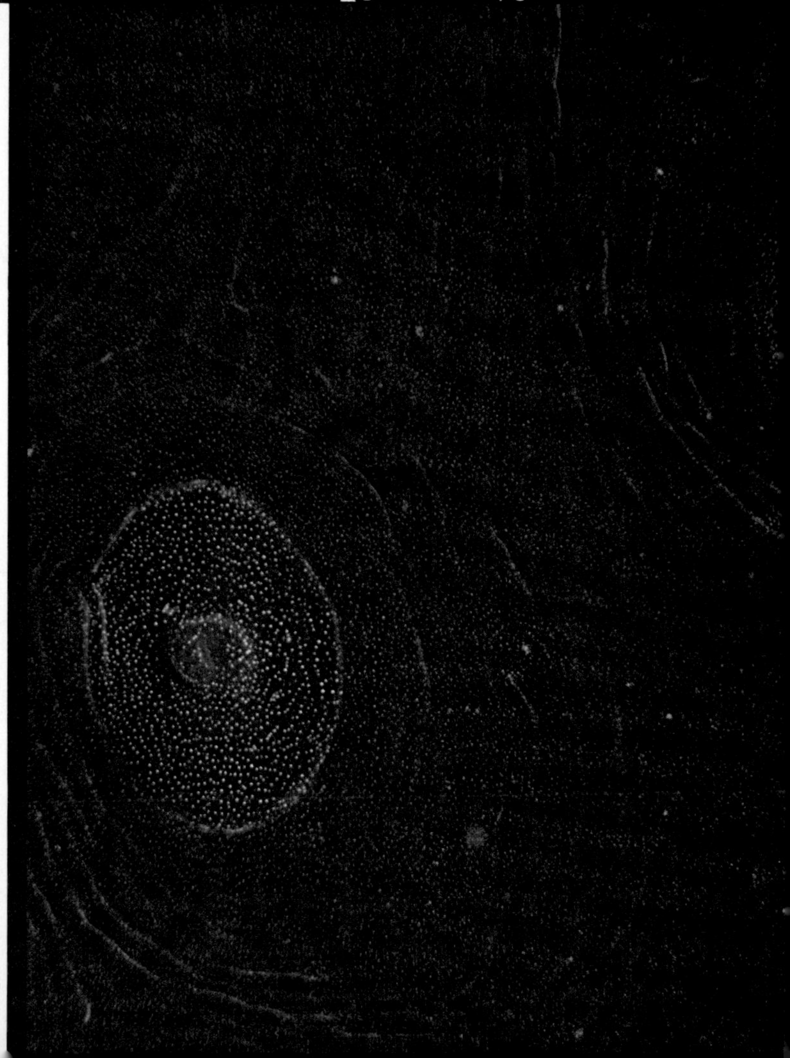

Chapter VIII. Photo Essay

The serene aquatic feeling of a blue weightless world is startingly transformed into a symphony of colors, patterns, and textures at close range. Artificial light reveals warm hues and enhances the abstract beauty of some of the more simple animals such as coral, sponges, and starfish.

***Tissues of tropical corals** (A, C, G) mask delicate skeletons to which they are comparable in beauty. The skeleton may resemble the tissue form (A) or appear totally different (C and G).*

*The not-so-spiny skin of some **starfish** is composed of plates, arranged in symmetrical designs (E) or randomly colored in pastels (B). The canals of a **yellow sponge** (D) exhibit abstract patterns.*

*Bearing no resemblance to its flowerlike form when expanded, this **anemone** (F) has contracted to a crimson blob on its rock.*

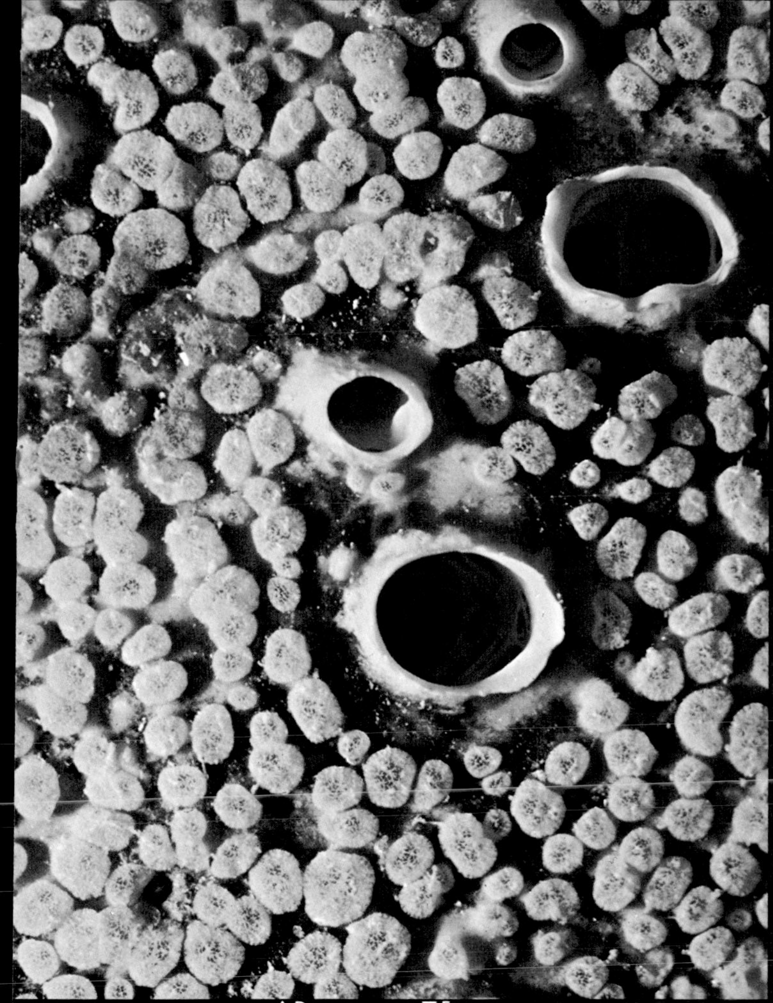

▲D

▲F

▼E

▼G

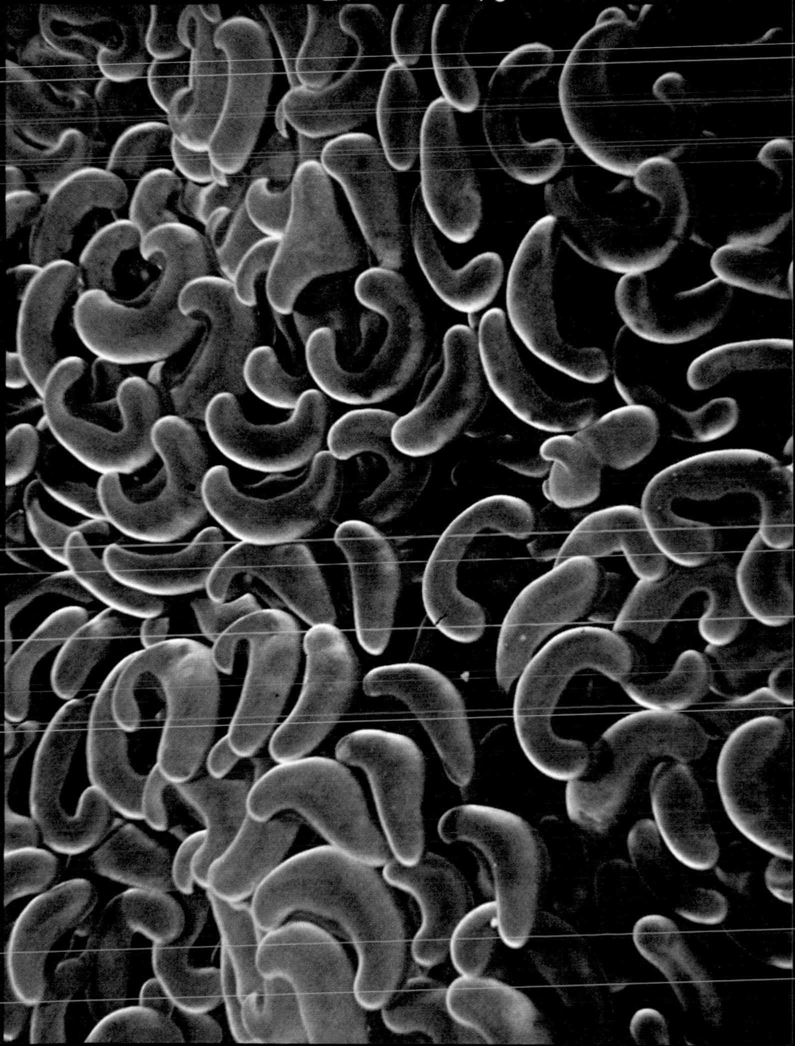

▼B ▲A

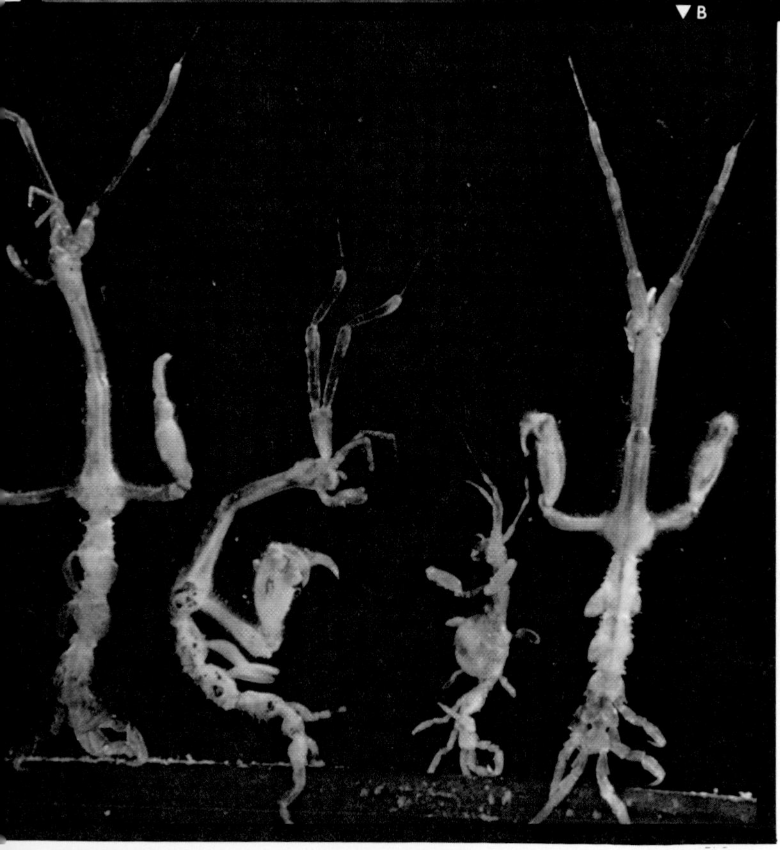

Insights into Beauty

An observant eye discovers some personalities of marine life and the moods they impart to a liquid world. Some appear as frivolous clowns, silently observed by expressionless, unblinking eyes. The passive idly sway in a gentle surge and are blind to danger, while the active may respond in panic-stricken flight.

***Flatfish** (A), thinking he is sufficiently camouflaged, observes with protruding eyes a camera's advance.*

***Caprellid shrimp** (B), idly bob up and down, as they wait for a planktonic meal to drift by.*

*The outstretched tentacles of a **tropical anemone** (C) have stinging cells that appear as texture.*

*A frightened **angelfish** flees (D) at the approach of a diver noisily spewing bubbles in his wake.*

C
D

▲A ▼B

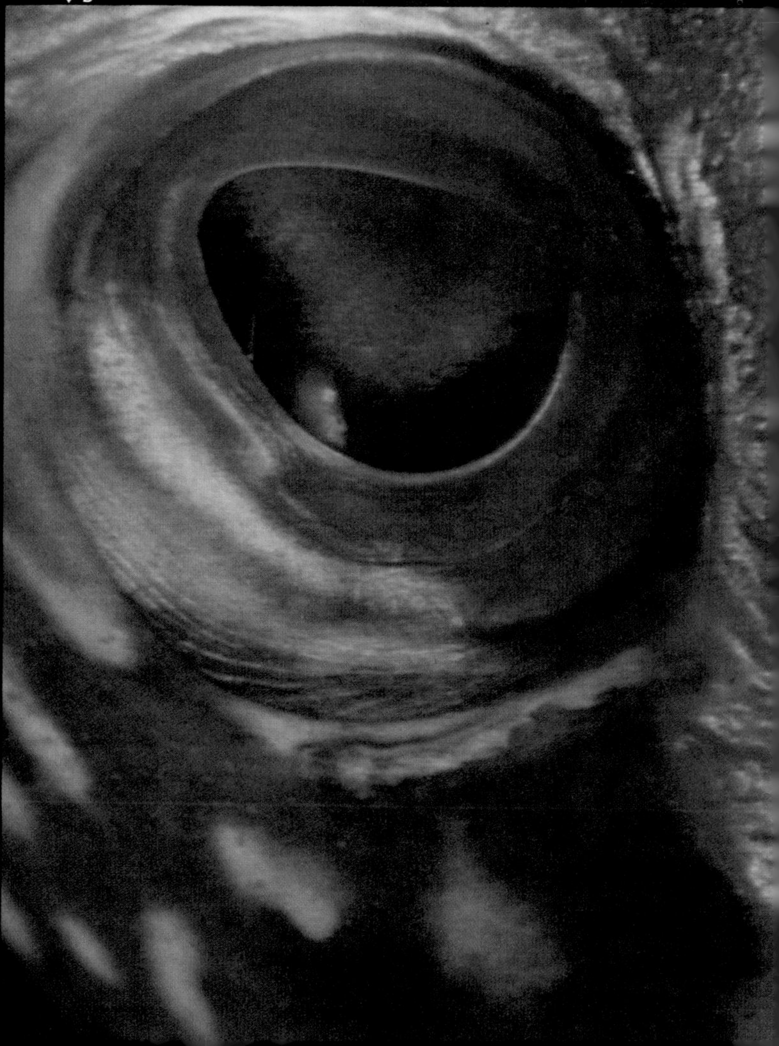

Strange Forms

Glazed eyes play hide and seek as they peer from ominous hidden forms, giving an impression of masks and make believe in the underwater scene. Strange dandelion shapes may add decoration to a sea fan or colorful animated forms may gracefully drift through a limitless blue space.

The iris of a ***rockfish eye*** *(B) exhibits a striking contrast in form to that of a well-camouflaged* ***flatfish*** *(A) lying on the bottom.*

Lacy hydroids *(C), supported by a sea fan, reach upward for a greater exposure to passing currents. This relationship is uncommon although sponges and anemones are often found living together.*

The delicate colors and bizarre appendages enhance the camouflage abilities of this ***South Pacific seahorse*** *(D) swimming in open water.*

C

D

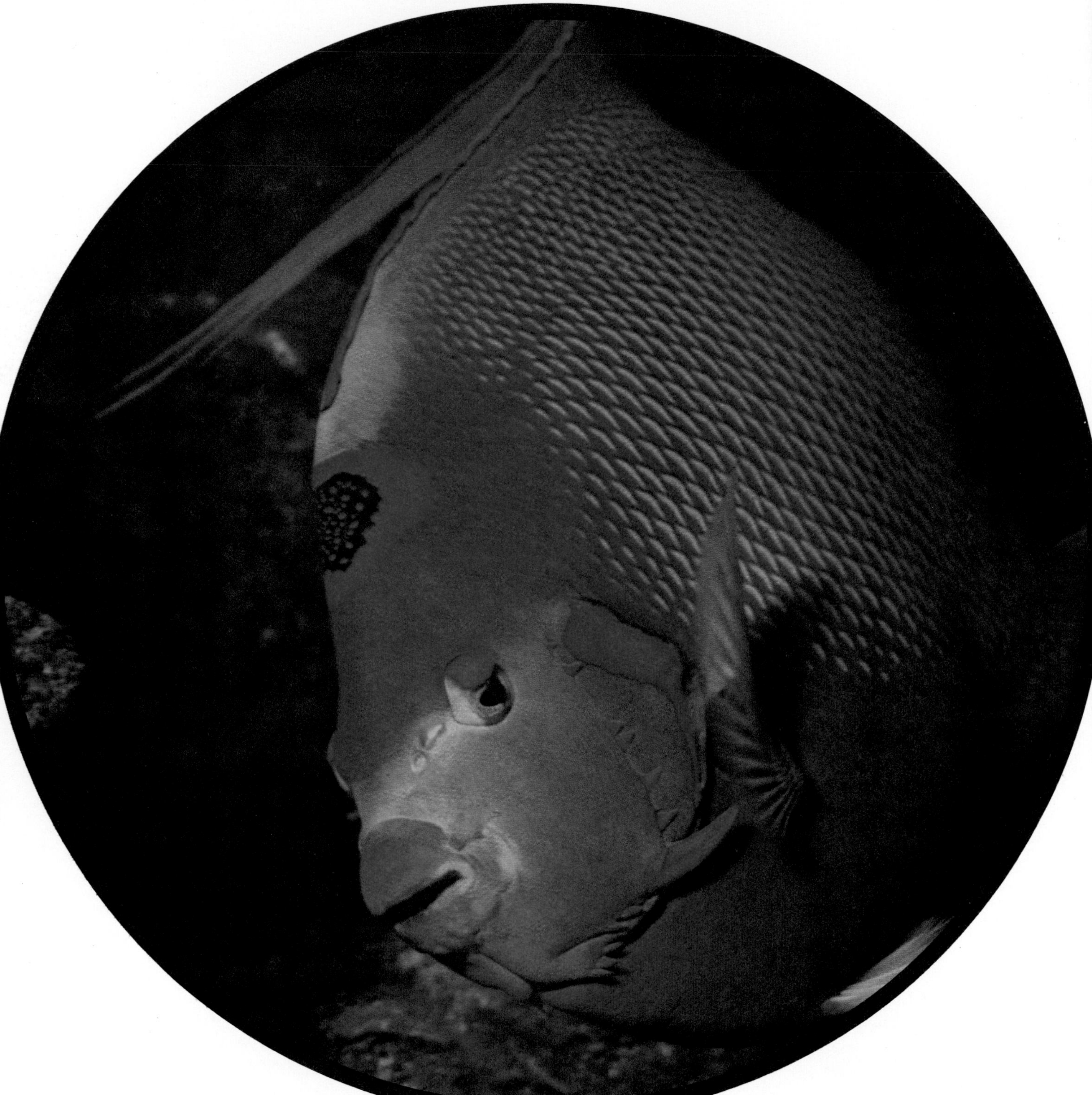

From Two Worlds

When aliens descend from an unknown world to the realm of angelfish, they stand out as ungainly intruders against a back-

drop of liquid beauty. The temporary visitors must behave cautiously to prevent destroying the very objects of their quest.

The seemingly defenseless ***queen angelfish*** *sports an effective spine on its gill cover (left). Appearing as* ***visitors from outer space,*** *our divers begin their descent to Conshelf II (above).*

Conclusion I. A World Without Witness...

"Eyes and ears are bad witnesses to men," said Heraclitus in the sixth century B.C., "if they have souls that understand not their language." Today, such a statement rings like an urgent warning: mankind, blindly engaged in the race of western civilizations, is in the process of severely devastating nature—and the oceans in particular—just at a time when the vital importance of the water cycle is clearly understood. As early as the sixth century B.C., Thales stated that, "The world is made of water." He meant to point out that food and beverages, as well as the blood and flesh of all living creatures, are basically aquatic.

Science deals with known facts; philosophy encompasses all sciences and generalizes further through speculation. Western culture encourages the proliferation and thus the ramification of art, technology, and sciences. In the process, the learned part of the population goes through a phase of overspecialization, using their own jargon made of new words that only the initiated can understand. The scientific aristocracy drifts away, physically and intellectually, from the average person. This social split is obviously extremely dangerous, and it is high time to bring science back to the people, who pay for it and own it.

The books in this series have scanned the eighteen most significant aspects of the multifaceted ocean world. In order to tell these stories in a simple manner and in a popular language, we went to the fountainhead of the marine sciences, to the treasury of facts and knowledge accumulated over hundreds of years by the heroic endeavors of many dedicated scientists from all countries. We have also relied on the personal observations of the *Calypso* diving team. We have interpreted the knowledge thus available in a spirit of love and of devotion for life inspired by Leonardo da Vinci's words: "Someday the slaughter of an animal will be punished as a murder," and by Stuart Mill's remark, "There is little satisfaction in contemplating a world without anything left to spontaneous activities of nature." The undersea world, like the rest of the universe, would hardly exist if there were no witnesses, anywhere, to realize the existence. Discovery is the next thing to creation. In discovering the marine world with us, you create it for yourself.

Jacques-Yves Cousteau

Grouper and Cleaners

Diver and Red Coral

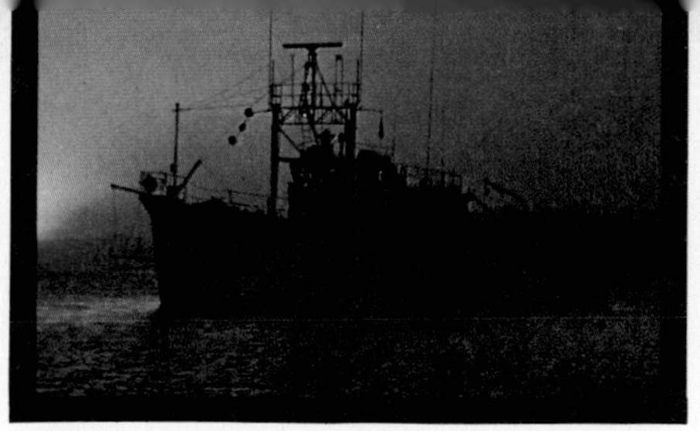
Calypso

Emperor Angelfish

Glossary

A

Abyss. The region of the ocean basin that lies below the continental slope, generally deeper than 6000 feet. It occupies three-quarters of the sea floor area.

Acanthodii. The most primitive group of jawed fish, often called "spiny sharks." They predate the placoderms but left a poor fossil record.

Actinopterygii. A subclass of the bony fishes, which includes most modern "normal-looking" fish. They have ray fins, in contrast to the crossopterygians which have lobe fins.

Adaptation. The physical, behavioral, and genetic changes a group of organisms undergoes to best succeed in its environment.

Adenosine triphosphate (ATP). An organic compound in living tissues that acts as an energy storage and transfer agent in metabolism.

Adsorption. The process of collecting substances, usually as liquids, on the surface of materials. This property is present in activated charcoal, certain clay minerals, and many other materials.

Aerosol. Particles suspended in a gas. In a sense, seawater suspended in air.

Agnatha. Fishlike creatures without jaws. This class of animals includes the lampreys, sea hags, and the extinct ostracoderms.

Alcyonarians. Members of a subclass of Anthozoans (corals and anemones) that includes the sea pens, sea fans, sea pansies, whip corals, and pipe corals. Alcyonarian corals are characterized by eight tentacles that are pinnate—that is, they possess featherlike side branches.

Algae. A large group of marine plants which contain chlorophyll but lack true roots, stems, and leaves. They may be single-celled or multicelled and may live alone or in colonies. Microscopic algae, called phytoplankton, are the base of the ocean's food web and carry out most of the photosynthesis that occurs in the sea.

Allantois. An embryonic membrane that forms in reptiles and birds, where it functions as bladder and lung. In mammals, the blood vessels in the allantois carry nutrient material from the placenta to the embryo.

Ambergris. A waxy, gray, strong-smelling substance that forms in the digestive tracts of sperm whales and is found in no other living creature. It is used as a perfume fixative.

Ammonoidea. A group of extinct cephalopods with shells commonly in flat spirals. Their name is derived from Ammon (Jupiter), a god who is often pictured with a spiraling ram's horn.

Amphibian. Any tetrapod vertebrate which spends portions of its life in and out of water and whose reproductive phase is entirely dependent upon a return to water. The larval forms usually have gills, which are not present in adults, as with frogs.

Amphipod. A member of the class Crustacea, order Amphipoda. They have no carapace, and their body is laterally compressed, giving them a shrimplike appearance. Amphipods range in size from one or two millimeters to the giant of the order, *Alicella gigantea*, 14 centimeters long. Most amphipods are marine, a few are semiterrestrial, and some are freshwater. Beach fleas are amphipods.

Ampullae of Lorenzini. Sensory organs in the head region of sharks and rays that are exposed to the exterior through small, round, porelike openings in the skin; they are very prevalent in the snout region. The function of the ampullae is not fully understood, but they are sensitive to changes in pressure, salinity, and electric fields.

Anaerobic. In the absence of oxygen; in contrast

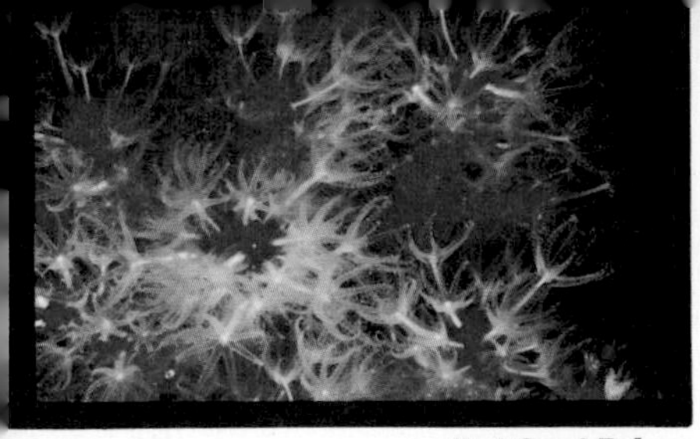

Red Coral Polyps

Rsch. Platform *(Isle Mysterious)*

Skuas

Squirrelfish

to *aerobic,* in the presence of oxygen. Both terms are commonly used in reference to respiration.

Annelid. The phylum that includes segmented worms possessing kidney and heartlike structures. Among the taxa are polychaetes, earthworms, feather-duster worms, and parchment worms.

Anoxia. A condition in which not enough oxygen is being carried by the blood to permit the cells of the body to carry on normal respiration. When prolonged, it can result in death. The cells of the nervous system and brain are the most sensitive to anoxia and may suffer permanent damage.

Anthozoa. The class of coelenterates that includes the anemones and corals. Anthozoans are exclusively polypoid and often have calcareous skeletons.

Anticline. A folded layer of rocks in which the center of the structure is elevated relative to the lateral sides.

Aqualung. The first automatic self-contained breathing apparatus for divers. It was coinvented by Emile Gagnan and Jacques-Yves Cousteau.

Arachnid. A class of arthropods that includes the spiders, mites, ticks, and scorpions.

Arthrodire. A group of placoderms which were rather efficient predators. A well-known member is *Dinichthys,* a 30-foot Devonian form.

Arthropod. The phylum of animals that includes the largest number of species. Arthropods have reached the peak of invertebrate evolution, exhibiting bilateral symmetry, a coelomic cavity, and well-developed organ systems—all characteristics of higher forms of life. They have segmented bodies covered with a hard exoskeleton and jointed appendages that give them a great deal of mobility. During the course of evolution they were the first great group to make the transition from land to water, where they have adapted to a wide range of niches. Insects, arachnids, and crustaceans are the best-known members of this phylum.

Ascidians. Members of the subphylum Urochordata, known as tunicates (sea squirts). They are common invertebrates found throughout the world and an important link in the evolutionary chain. Their larvae are bilateral and have a notochord; thus they are precursors of the vertebrate body plan.

Asexual reproduction. A form of reproduction that requires a single parent; often the mode used by simple invertebrates. The techniques include budding and fission.

Asthenosphere. The upper, plastic portion of the earth's mantle. It is the region where convective motions take place.

Atoll. A ring-shaped coral reef, with an enclosed lagoon, formed by the subsidence of a volcanic island around which the coral reef had originally formed. Carbonate sand may fill in some of the shallow reef areas, forming atoll islands.

ATP. *See* Adenosine triphosphate

Aurora australis. The Southern Lights. An atmospheric phenomenon resulting in bright bands of light flashing across the sky in the Southern Hemisphere. The *Aurora borealis* is a similar phenomenon in the Northern Hemisphere.

Autotrophs. Organisms that manufacture their own food, usually by photosynthesis.

B

Bacteriophage. A virus or possibly an unknown microorganism that inhabits and destroys bacteria.

Balance of nature. The equilibrium between organisms, including predators and prey, that can be damaged or destroyed by the removal of one or more animals or plants in an ecosystem.

Baleen. The long flexible plates made of material similar to that of human fingernails that hang down from the upper jaw of many species of whales, including the rorqual and blue. Whales swim with their mouths open, and the baleen functions as a strainer, separating krill and other microscopic animals from seawater.

Barbels. Small feelerlike extensions of the underjaws of certain fish, such as cod. They contain sensory receptors for taste (or smell) and touch.

Bathyscaphe. A free-diving device that functions in water much as a dirigible balloon does in air. It consists of a large cylindrical tank, filled with an incompressible lighter-than-water liquid, to which a passenger observation compartment is attached. Bathyscaphe means "deep boat" in Greek.

Bathythermograph (BT). A device used by physical oceanographers to measure the temperature of ocean water as a function of depth. As the instrument descends, temperature as related to depth is automatically plotted on a smoked-glass slide within the BT.

Benthic. Pertaining to the sea bottom. Animals inhabiting this region are termed *benthos*. Also appears as *benthonic*.

Berm. A feature characteristic of most beaches. The berm is the backshore part of the beach that runs from the sand cliffs or dunes seaward to the berm crest, from where the beach slopes more drastically to the water's edge.

Binary fission. A means of asexual reproduction in which the parent organism duplicates its components and then splits into two daughter cells.

Bioconcentration. The process by which predatory animals concentrate substances derived from the tissues of their prey. Many contaminants are concentrated via the food chains in this fashion.

Bioluminescence. The emission of cold light by living organisms; common in deep-sea creatures. The source of light is either certain associated bacteria or photophores (luminescent cells) in the animals. Used by organisms to differentiate species and sexes, predators and prey, in a world where no sunlight penetrates.

Biosphere. The living world; all the life on earth. This encompasses all the habitable regions of the earth including those in the atmosphere, hydrosphere, and the like.

Bivalve. Refers to molluscs with two shells. Clams, mussels, scallops, oysters, and cockles are examples. Formerly called pelecypods or lamellibranchs.

Blowhole. The hole in a cetacean's head through which it breathes and its spout passes.

Blubber. The thick layer of insulating fat found on warm-blooded marine animals.

Blue holes. Underwater caverns in which the water is a royal blue due to the conspicuous absence of animal and plant life and the depth of the water. Blue holes were formed above sea level by the erosive action of water thousands of years ago. When the glaciers melted and sea level rose, these caverns were flooded with seawater.

Brachiopod. Any member of the phylum of bivalved animals that closely resemble bivalve molluscs in outward appearance, but that are more primitive in organization. Shells are dorso-ventrally oriented. They are commonly found as fossils. *Lingula*, a living brachiopod, dates back to the Cambrian period, some 570 million years ago.

Brownian motion. The constant zigzag movement of colloidal particles suspended in a liquid; caused by the collision of the particles with rapidily moving molecules of the liquid medium. This phenomenon is named after Robert Brown (1773-1858), who first described it.

Bryozoans. A phylum of animals commonly called "moss animals" (from the Greek *bryon*, or moss). They are microscopic and form colonies that resemble thin, matlike, algaelike growths on shells, rocks, etc. Colonies can gain considerable size; some float free, often confused with seaweed. They are very important as fossils of the Paleozic era.

Budding. A mechanism of asexual reproduction in which the parent creates one to several daughters as appendages. These daughter cells can detach themselves and exist separately; some, however, may remain attached and thereby produce colonies.

Buffer. In chemistry, a substance that aborbs excess acidic or basic ions to maintain a balanced pH in a solution.

Buoyancy. The property of being suspended by a difference in density with the surrounding medium.

Buttress zone. The outer (seaward) limit of a reef system. The base generally slants steeply and forms a "wall" against which the waves crash.

Byssus. Strong, threadlike material secreted by some marine creatures (mussels and some other bivalves) that helps them adhere to submerged surfaces. The fabled golden fleece.

C

Calcification. The process of adding calcium to a material. It may result in the stiffening of a skeleton or shell or in the creation of a calcium framework.

Carapace. A hard protective shell that covers part, or all, of the upper surface of an animal. Common in arthropods; we speak of the carapace of the lobster or crab.

Carbonization. One of the fossilizing processes which occurs due to the distillation of organic substances in the tissues, resulting in a carbon film.

Carbon cycle. The ongoing cycle of carbon in nature: the intake of carbon dioxide by plants; its conversion to sugars and subsequent metabolism by plants and animals; and finally the return of carbon dioxide to the air through respiration.

Carnivore. An organism that eats the tissues or cells of animals.

Cartilaginous fish. A fish which lacks bony structures. The internal skeleton is composed of cartilage instead of bone. Examples include sharks, rays, and many extinct fish.

Cell. The basic organizational unit of life. It is composed of organelles or undifferentiated protoplasm and is usually bounded by a membrane or wall.

Cephalopod. Any of a class of molluscs with tentacles extending from the head. They are usually free-swimming and have well-developed eyes. Shell may be internal or external. Examples are octopods, squids, cuttlefish, and nautiloids.

Cephalotoxin. Poisonous substances found in the beaks of several cephalopods, notably the octopus.

Cetacean. Any of an order of aquatic, mostly marine, mammals (whale, dolphin, and porpoise). Appendages reduced, tail expanded into flukes.

Chlorophyll. The green pigment found in photoautotrophic organisms that catalyzes the photosynthetic process.

Chondrichthyes. *See* Cartilaginous fish.

Chondrostean. A fish belonging to the most primitive group of the ray-finned fish. The sturgeon is a living example.

Chordate. An animal with a notochord or vertebral column, gill slits at some stage of its life, and a dorsal hollow nerve cord. Mammals, birds, fish, and *Amphixous* are all chordates.

Chorion. One of the embryonic membranes surrounding a developing fetus; its function is protective and nutritive. In mammals with internal development, the chorion forms part of the placental connection with the mother's tissue, through which the embryo receives nourishment.

Chromatophore. A cell in a fish or other organism that contains a colored substance. By controlling the distribution of pigment in its chromatophores, a fish can change its color patterns and camouflage itself against a variety of different backgrounds.

Chromosome. A threadlike strand within the nucleus of a cell that contains the genes; the bearers of the genetic code, governing all aspects of cellular development and reproduction. Chromosomes are constant in number within a given species. (*See* DNA.)

Cnidaria. *See* Coelenterate.

Coacervate. A concentrate of molecules in suspension, held together by ionic forces; usually formed on a solid surface such as clay.

Coccolithophore. A tiny, spherical, planktonic plant covered with calcareous plates. Coccoliths (the plates) are one of the major sources of calcareous sediments on the ocean floor.

Coelacanth. A member of a group of lobe-finned fishes that may be ancestors of the land vertebrates. This group was thought to have been extinct since the Mesozoic era, but several specimens of the genus *Latimeria* have been caught during the past 40 years off Madagascar.

Coelenterate. A member of the phylum Cnidaria, (often called Coelenterata); characterized by a body cavity with both digestive and circulatory functions, by stinging cells (nematocysts), and two distinct tissue layers. Examples of coelenterates are corals, jellyfish, and hydroids.

Coelom. An embryonic cavity from which the other body cavities are derived; found in most multicellular organisms. In adult animals, the coelom houses the digestive tract and various other organs associated with it.

Commensalism. A relationship between animals and/or plants, in which there are advantages for some of the partners while none are adversely affected. Pilot fish and sharks are an example.

Continental drift. The theory currently in favor, that the continents on earth have drifted into their current positions after once having together formed one or more landmasses.

Continental shelf. The portion of the sea floor that extends from the low-tide mark to the abrupt break in slope at a depth of approximately 600 feet (200 meters). The seaward edges of the shelves mark the true continental boundaries.

Continental slope. The steep slope that extends from the continental shelf seaward to the abyss. In some places it is broken by the *continental rise* (a change in slope or actual elevation) before it reaches the sea bottom.

Convergent evolution. Similarity in form occurring in two separate evolutionary lineages, such as in the tuna and mako shark.

Copepod. Any of a large subclass of small freshwater and marine crustaceans that are a major constituent of zooplankton.

Coral. A solitary or colonial anthozoan coelenterate that characteristically secretes a calcareous skeleton. These skeletons may form *coral reefs.*

Countershading. Color patterns exhibited by fish and cetaceans. These animals are darker on top and lighter below; thus they appear an even gray when strong sunlight shines from above.

Crinoid. Stalked pelmatozoan echinoderms, very common as fossils from the Carboniferous period. Surviving crinoids are referred to today as "sea lilies"—flower-shaped marine animals that are anchored to the substrate by a stalk opposite the mouth, common on reefs in the Indo-Pacific.

Crossopterygii. Lobe-finned fishes, including the coelacanths, that may have been the ancestors of the land vertebrates.

Crustacean. Any member of the largest group of marine and freshwater arthropods, including such diverse forms as lobster, copepods, and barnacles.

Current. Movement in water masses caused by density differences, winds, the Coriolis force, and other energy sources.

Cytoplasm. The material that makes up the living cell. A specialized form of *protoplasm* (which is material that makes up any living substance).

D

Deep scattering layers. A series of layers formed by certain marine organisms that reflect or scatter sound. Most of these organisms rise toward the surface at night and may range as far down as 2000 feet in the daytime. Some layers are unaffected by the diurnal cycle.

Detritus. In the biological sense, organic particles that settle to the sea bottom, where some types of benthic organisms can use them as a source of food.

Diatom. A single-celled planktonic plant of the subphylum Chrysophyta. They are characterized by a siliceous test and are among the most abundant of all marine phytoplankton.

Dinoflagellate. A single-celled planktonic plant of the subphylum Pyrrophyta. They are characterized by two flagella that they use for locomotion. When environmental conditions are favorable, explosive population increases occur, causing in some cases a dangerous red tide that can result in extensive fish kills and the contamination of shellfish in the area. Many dinoflagellates are bioluminescent.

Divergent evolution. Evolution in very different directions from a common ancestor, resulting in closely related forms looking very different. A good example is the dissimilar appearance of sea urchins and sand dollars.

DNA (deoxyribonucleic acid). The complex organic compound in the genes that controls the replication and activities of living tissue. These helical molecules, along with certain proteins, constitute chromosomes. *RNA* (ribonucleic acid) acts as a messenger in conjuction with DNA.

Doldrums. A region of calm winds along the equator, caused by the rising of warm air in that area.

Dolphin. Cetaceans of the family Delphinidae, generally possessing a beaklike snout and numerous teeth. They are not to be confused with the fish also called dolphins.

Doppler effect. The apparent change in frequency of sound or light waves caused by differences in the direction and velocity of the sound or light source and of the observer.

E

Echinoderm. Any of a phylum of radially symmetrically, coelomate marine animals, generally having a high level of organization. Although adults are radially symmetrical, larvae are bilateral. They are an important step forward in the course of evolution, as all higher forms of life are bilateral. This phylum includes the starfish, sea urchin, and sea cucumber.

Echo sounding. A technique of determining the depth of the sea bottom by the use of sound reflection; the results are produced graphically *(echo-*

gram). The device used is termed an *echo sounder, depth sounder,* or *depth reader.* A similar technique for locating objects is called *echolocation.*

Ecosystem. The totality of organisms in a community; sometimes refers to physical attributes of an organic community. It may be considered in equilibrium if the community is stable for a finite period.

Electrolyte. A liquid solution capable of conducting electric currents because its dissolved components can dissociate into their ionic form. Salt ($NaCl$) in ionized form (dissolved—Na^+ and Cl^-) is a typical example.

Elver. A young eel recently metamorphosed from its larval stage.

Embryology. The science dealing with the earliest development of an organism.

Empiricism. A search for knowledge by observation; sometimes used in contrast to *experimentalism* (the pursuit of knowledge through direct testing in controlled conditions).

Enzyme. An organic catalyst; something that affects the rate of a reaction without being permanently altered itself.

Epifaunal organisms. *See* Infaunal organisms.

Epoch. A subdivision of time, commonly used within the Tertiary period of the earth's history.

Equilibrium. A state of balance; in biology the steady state of undisturbed interactions; in chemistry a reversible reaction remaining unchanged.

Era. The largest subdivision of earth time, e.g., Paleozoic, Mesozoic, Cenozoic.

Estuary. A semiclosed body of water with free access to the open ocean; wherein seawater is measurably diluted by freshwater runoff from land and where the tides of the ocean affect water movements. Most estuaries are drowned river valleys or valleys cut by glaciers (fjords).

Eucaryote. An organism, one- or many-celled, that has its genetic material bound up in a nuclear membrane. This is in contrast to *procaryotes* which have their genetic material dispersed through their cytoplasm. Procaryotes include bacteria and blue-green algae. All higher forms of life are eucaryotes.

Euphausiids. Pelagic, shrimplike crustaceans averaging one inch in length. All members of the order Euphausiacea are marine, and most are filter-feeders; a few are predaceous. An important part of the polar food chain; whalers called them "krill."

Euryterid. One of the group of extinct "sea scorpions" that lived in the early Paleozoic era.

Eutrophication. The natural aging process by which the dissolved oxygen in a standing water body, usually a lake, is reduced to the point where most forms of life can't survive.

Evolution. A gradual process of change. Usually used in the sense of "organic evolution," which is the change in living organisms, often in the direction of specialization.

Extinction. The destruction or termination of something; usually refers to a group of organisms.

F

Fathometer. A sonar device used for measuring ocean depth; also called a *depth sounder.*

Fault. A fracture in the earth that involves movement of the rocks on either side of it.

Fertilization. Initiating the development of offspring by means of the introduction of genetic material in sexual reproduction; combining of an egg and sperm. It may be *external* (occurring in the water) or *internal* (occurring within the mother).

Flagellum. A threadlike or whiplike extension of the protoplasm of some cells, often used in locomotion such as the tail of a sperm cell.

Flagellata. A group of flagellate protozoa in the subphylum Mastigophora.

Flukes. The caudal extensions of a cetacean; they form the forked, horizontal fishlike tail.

Fluorescence. The property of emitting light under stimulation by an outside energy source such as X or ultraviolet rays.

Food chain. The succession of predation from the phytoplankton, to the zooplankton, to small fish and larvae through various levels, to the largest predators such as tuna and man.

Foraminifera. Any of an order of marine protozoans with chambered shells; so named because the shells usually have holes (or foramina) through which pseudopods extend. They are abundant planktonic and benthonic animals.

Fossil. Any preserved remain or trace of a dead organism; generally older than the Recent period.

Fossil fuels. The energy sources man uses which derive from fossil organisms, including coal, natural gas, and petroleum.

G

Gamete. The reproductive cell of sexually reproducing organisms, possessing one-half the full number of chromosomes (haploid).

Gastropod. Any of a large class of molluscs, often with a univalve shell. Examples are snails, nudibranchs, and limpets.

Gene. A replicative unit in the chromosome, composed of DNA.

Geotaxis. An innate response by a living organism to the gravitational forces of the earth.

Gestation. The period of time in the development of an individual from conception to birth.

Gill. The complex of breathing organs in an aquatic animal, composed of osmotic membranes, supports, clefts, and vascular tissue.

Glacier. A mass of compacted snow and ice that forms when the rate of melting is less than the rate of snow accumulation.

Gonad. The reproduction organ of an animal, usually the ovary or testis.

Gondwanaland. The supercontinent that may have once incorporated all the southern landmasses before continental drift took place more than 60 million years ago.

Graptolite. An extinct protochordate planktonic animal, possibly related to the living pterobranch.

Great Barrier Reef. A 1250-mile-long coral reef off the coast of Australia; the largest coral reef in existence today.

Greenhouse effect. The superabundance of atmospheric carbon dioxide over many industrialized areas mimics the glass of a greenhouse and acts as a heat trap. This phenomenon may someday warm the earth enough to melt the ice caps and precipitate a glacial advance or "ice age."

Guano. The feces of birds and bats, commercially important as fertilizer.

Gulf Stream. The warm ocean current, approximately 50 miles wide, flowing from the Gulf of Mexico along the eastern U.S. coast, then turning toward Europe at the Grand Banks.

Guyot. A flat-topped submarine peak. Guyots are indicative of past, lower sea levels and/or submergence of the seamount.

H

Hadal zone. The oceanic trenches.

Herbivore. An organism that subsists on plant nutrients; a plant-feeder.

Hermaphroditic. Having the sex organs or reproductive aspects of both male and female. Many invertebrates and some fish show this capability.

Heterotroph. An organism that must depend on food outside itself; usually a form that directly or indirectly eats plants. Opposed to *autotroph.*

Holograph. A photograph taken with laser light to produce three-dimensional images. The process requires no lenses and works by interference phenomena in the split laser beams.

Holoplankton. Members of the planktonic community that spend their entire life cycles as part of the plankton. This is in contrast to *meroplankton:* planktonic forms that spend only part of their life as plankton. Examples are fish eggs and larvae. Nanoplankton is a size classification: very small forms, ranging from 5 to 60 microns. (One micron $= 10^{-4}$ cm or 3.94 x 10^{-5} inches.)

Holosteans. Any of largely extinct group of fish characteristic of the Mesozoic era. Living members include the gar pike and bowfin.

Holothurians. A class of echinoderms known as sea cucumbers. They are bottom dwellers, living in sand and mud. They have retractable tentacles near their mouth with which they capture food.

Hydrodynamics. The branch of physics dealing with the motion and forces of water.

Hydroid. A polypoid form of coelenterate in form like the genus *Hydra.*

Hydrophone. A device for detecting and pinpointing underwater sounds.

I

Ichthyologist. A specialist in the study of fishes, their anatomy, classification, and life history.

Igneous rock. A rock that has solidified from molten magma, either within the earth or on the surface in the form of volcanic lava.

Inbreeding. The process of continually mating individuals from closely related or identical ancestors.

Infaunal organisms. Animals that live in the ground within a certain area. We speak of the infaunal organisms of a mudflat, etc. This is in contrast to *epifaunal organisms* that live above ground.

Instinct. An inborn tendency to behave in a certain way which is characteristic of a species.

Internal wave. An ocean wave generated along a horizontal density gradient between two water masses beneath the surface. Unlike wind-generated waves, they are not visible on the surface. Internal waves can be a hazard to submarines.

Intertidal zone. The portion of the sea and sea bottom included between the low- and high-tidal marks. Most obviously seen on rocky coasts.

Invertebrates. Those animals that lack a spinal column or notochord; that is, all animals except fishes, amphibians, reptiles, birds, and mammals.

Isopods. Any of the second largest order of crustaceans next to the Decapoda (lobsters, crabs, shrimps). Most isopods are marine, a few are freshwater, some are terrestrial, and a few are even parasitic. They have a characteristically flat body and lack a carapace. Most isopods are 5 to 15 mm long. The giant of the group is a 14-inch isopod that lives in the deep sea *(Bathynomus giganteus)*.

Isostasy. Approximate balance of areas of the earth's crust which tend to "float" on the semifluid mantle below. Continents consist of material lighter than that of the ocean floor and thus they float higher than the sea floor.

J

Jacklighting. Drawing fish to a net or fishing area by use of torches or lights.

K

Kelp. Common name for large brown seaweeds belonging to the brown algae.

Kraken. A Norwegian sea god and name for giant squid or octopods.

L

Labyrinthodont. A long-extinct group of primitive amphibians. Fossil labyrinthodonts are common in deposits from the Carboniferous period, which are 300 million years old.

Larvae. The immature form of any animal that changes structure (usually by metamorphosis) as it becomes an adult. Usually free-swimming.

Lateral-line organ. A series of sensitive hairlike cells imbedded in a matrix that respond to changes in pressure. These sensory receptors run laterally down the body of many fishes. They can detect water pressure, direction and rate of flow of water, and low-frequency sound.

Laurasia. A hypothetical continent composed of North America and Eurasia that supposedly separated in the late Paleozoic era.

Lithosphere. The solid portion of the earth; the earth's crust, as opposed to the atmosphere.

Littoral. Referring to the shore or coast; the environment of flora and fauna from the shore to a depth of about 600 feet (200 meters).

Loran. A navigation system by which a ship can determine its position by the difference in time between radio signals from two or more known locations. Other systems include DECCA and Omega.

Lysosomes. Cellular organelles, saclike in shape, enclosing enzymes from the rest of the cytoplasm. The enzymes, thus contained, break down cellular macromolecules, or the organism itself after death.

M

Mammal. Any of a class of vertebrates usually characterized by homeostatic body temperature, live birth, and mammary glands.

Mantle. Layer of earth between the *crust* (lithosphere) and *core;* also, the fleshy fold of tissue that secretes the shell of the mollusc.

Mariculture. The technique of artificially cultivating and culturing marine plants and animals for human consumption. It is most successful, at this point, with shellfish.

Meiosis. Cell division in which the chromosome number in the resulting cells is haploid, or half the usual number (reduction division). Meiosis results in the formation of cells for sexual reproduction; when two sex cells unite, the normal diploid number of chromosomes is restored. This is the opposite of *mitosis,* in which the resulting cell is diploid. Mitosis is the process that almost all the cells of our body use to replicate themselves.

Melanophore. A specialized type of chromatophore responsible for dark pigmentation; found in ciphalopods, crustaceans, and fish.

Mesoglea. Jelly-containing layer of tissue between the ectoderm and endoderm of coelenterates.

Metabolism. Processes by which food is assimilated into tissue (anabolism) and those by which accumulated matter is broken down (catabolism) to release energy in living organisms.

Metamorphosis. Process by which an immature form of an animal undergoes radical structural change into the adult form.

Metazoans. Term used in some systems of taxonomy to refer to all animals whose bodies are composed of complex, differentiated cells.

Mid-Atlantic Ridge. Great ridge-and-valley system rising in the center of the Atlantic, running north and south throughout the ocean. One of several oceanic ridge systems found throughout the world, it is the site of sea floor spreading.

Migration. Instinctive response to reproductive or food needs that triggers repeated movement of entire species to a new area. *Vertical migration* refers to the diurnal movement of plankton from deep to shallow depths in the sea.

Mimicry. The close resemblance in behavior, coloration, or physical appearance of one organism to another. It usually is protection from predation.

Mitochrondia. Cytoplasmic organelle that serves as an enzyme storehouse and is vital to the metabolism of the cell.

Mitosis. See *Meiosis.*

Mohorovicic Discontinuity. Whenever there is an earthquake or nuclear blast, for example, shock waves travel through the earth. When these waves meet areas of different densities, they speed up or slow down. Waves abruptly increase in speed below 50 km, indicating a change in material. This area of change became known as the Mohorovicic discontinuity, named after the Yugoslavian seismologist who discovered it. Called Moho for short, it is the boundary between the earth's crust and mantle.

Molecule. Smallest particle of substance or compound that can exist in the free state while still maintaining the properties of the material.

Molluscs. Any member of the phylum Mollusca; generally characterized by a soft, nonsegmented body with gills, enclosed in a mantle and shell.

Mouthbreeders. Any of the fresh and salt water species of fishes that take fertilized eggs into the mouth for incubation. Common marine varieties include the catfish, cardinalfish, and jawfish.

Mutation. Any change, artifically or naturally induced, in the genetic makeup of an animal that will be incorporated into the chromosomal material of the succeeding generation.

Mutualism. A symbiotic relationship between two species of organisms that is mutually beneficial.

N

Nacre. The pearly inner layer in the shell of several molluscs. Often called "mother-of-pearl," it has been used for centuries as an ornamental material.

Nautiloids. Any of the subgroup of cephalopods; the only living member is the nautilus.

Nekton. All free-swimming organisms whose movements are not governed by currents and tides.

Nematocysts. Stinging cells of coelenterates.

Nephridia. Kidneylike structures that perform excretory functions in many invertebrate groups. The marine worm *Nereis* has a pair of nephridia in each of its body segments.

Neurons. Nerve cells; the structural and functional units of the vertebrate nervous system. The squid, an invertebrtate, has a giant neuron, which makes it a valuable animal to research scientists.

Nitrogen narcosis. A mental disorder resulting from breathing compressed air below 100 feet. Often called "rapture of the deep," it results in double images, exaggeration of some senses and attenuation of others, feelings of euphoria and terror, and poor mental judgment. A sufficient explanation for this disorder has not been found, but it is related to the presence of nitrogen in the circulatory system.

Notochord. An elongated rod that acts as a supporting structure in lower chordates. It appears in the embryonic stages of higher vertebrates but is later replaced by the vertebral column becoming the centrum of the vertebrae.

Nototheniidae. A group of deep-sea fishes with worldwide distribution.

Nucleic acids. A group of acids that carry genetic information. The two kinds are DNA and RNA. They can be found in various cell structures as well as in the cytoplasm. (*See* DNA.)

O

Olfactory receptors. Sensory elements that detect chemicals dissolved in water or in air. Touch is not involved, as it is in taste.

Ontogeny. The development of an individual.

Ooze. A general name given to sediment deposits on the floor of the deep ocean that were derived from biological sources.

Osteichthyes. In some taxonomies of living things, the class to which all bony fishes belong. The other large class of fishes surviving today is the Chondrichthyes, or cartilaginous fishes.

Ostraciiform movement. A type of locomotion in fish where the body remains rigid and only the tail fin flexes, providing all the forward propulsion. It is used by boxfish and trunkfish.

Ostracoderm. Primitive fish that roamed the seas 400 million years ago. They lacked jaws, and their bodies were covered by thick bone or bonelike plates. Ostracoderm means shell-skinned.

Outfall. Discharge from treatment plants, factories, power plants, and the like that is released into the aquatic environment.

Oviparous. *See* Viviparous.

Ovoviviparous. *See* Viviparous.

Oxidation. The combination of a substance with elemental oxygen. In more general terms, a substance is said to be oxidized if it loses one or more electrons in a chemical reaction.

P

Pangaea. The supercontinent believed to have existed 200 million years ago. Pangaea broke up, and the pieces began to drift apart, forming the continents as we know them today.

Parasitism. A relationship in which one organism obtains food and other benefits at the expense of the other; the host organism is usually harmed.

Parthenogenesis. Unisexual reproduction in which the egg begins to develop and grow without ever having been penetrated by a sperm cell or accepting its nuclear material.

Pelagic zone. All ocean waters covering the deep-sea benthic province (bottom). The pelagic zone is divided into an open sea (oceanic) province and an inshore (neritic) province.

Permafrost. Permanently frozen subsoil found in the higher latitudes of the earth.

Petrification. A process in which organic matter is replaced by silica, lime, or some other mineral deposit to form a stony substance.

Pheromone. A substance secreted by an organism that stimulates a behavior or physiological response in another individual of the same species.

Phosphorescence. The emission of visible light without the production of any noticeable heat. This is a common phenomenon among deep-sea creatures. A synonym for this term in is bioluminescence.

Photoautotrophs. Organisms that depend upon light as an energy source (for photosynthesis) and which use carbon dioxide as their principle source of carbon.

Photon. A unit, or quantum, of electromagnetic energy. The energy of light is carried by photons.

Photosynthesis. A biological process in which plants convert carbon dioxide and water into usable carbohydrates. Solar energy, captured by the green pigment chlorophyll, supplies the power to carry out the many complex chemical reactions that are involved in the photosynthetic process.

Phylogeny. The evolutionary development, or lines of descent, of a plant or animal species.

Phytoplankton. The microscopic plants that inhabit the oceans of the world. They are responsible for most of the photosynthesis that goes on in the ocean and are the basis of the oceanic food web.

Pingos. Ice intrusions in arctic waters that resemble reef formations.

Pinnepeds. A group of marine carnivores that includes all the seals and walruses. Characteristically the digits at the end of each limb are connected and covered with a thick web of skin. Most species possess claws.

Placenta. A structural link between developing embryo and mother through which the embryo receives nourishment and rids itself of waste products. It is a mammalian characteristic.

Placoderms. Any of a group of archaic jawed fishes that became extinct by the end of the Paleozoic era, about 200 million years ago. Placoderms, like ostracoderms, were covered, to a degree, by armor plates.

Plankton. In general, any organism, large or small, that floats or drifts with the movements of the sea. Plants are called phytoplankton, and animals zooplankton. Most organisms that spend their entire life as part of the plankton are microscopic in size.

Plate tectonics. The study of the earth's crustal plates and the forces that cause them to "drift" over denser mantle rocks.

Platyhelminthes. The phylum to which the flatworms belong. Two classes, the Cestoda (flukes) and Trematoda (tapeworms), are parasitic. A third class, Turbellaria, is free-living. The flatworms are the most primitive bilaterally symmetric animals and are an important evolutionary link.

Polymerization. A process by which two or more molecules are united to form a more complex molecule. The new molecule has different physical and chemical properties from those of its components.

Polyp. The body form of various coelenterates that remain attached to the substrate. A polyp has a hydralike form; it is a fleshy stalk with tentacles. Polyps can occur singly or as part of a colony. The polyp stage of a coelenterate's life cycle may be contrasted to the medusa or free-swimming stage.

Preadaptation. A quality possessed by a living organism that is not essential for its survival today but might be vitally important to its existence at some future stage of development.

Predator. A creature that survives by capturing and feeding on other living organisms.

Pressure drag. A resistance to forward motion that results from turbulent forces generated as an object or animal moves through its medium. A fish with a fusiform shape produces little turbulence and therefore reduces the pressure drag holding it back.

Primary production. The amount of organic material (represented by carbon) produced per given unit of seawater by photosynthetic autotrophs.

Procaryotic cell. A cell that lacks a membrane-bound nucleus. Nuclear materials are found within the cytoplasm.

Protochordates. A group of primitive animals possessing a notochord during some part of their life cycle (acorn worms, pterobranchs, and tunicates).

Protozoa. Microscopic, nonphotosynthetic, eucaryotic, unicellular animals. Major groups include the ciliates, amoeba, flagellates, and sporozoans.

Pterobranchs. Tiny marine animals that form plantlike colonies. They are considered the oldest protochordates as they show primitive gill development and a food-filtering system from which more typical chordate features could have been derived.

Pycnogonid. A deep-sea arthropod commonly known as a sea spider. Almost all legs and no body, these unique creatures have successfully adapted, through evolutionary processes, to life in the abyss.

R

Radioactivity. A phenomenon exhibited by certain elements wherein various forms of radiation are emitted as a result of changes in the nuclei of atoms.

Radiolaria. One-celled marine animals (Protozoa) that are encased in a spiny coat of silica.

Radula. A tonguelike structure found in many molluscs. It has a rough surface and is used to rasp or file organic material from the surfaces of rocks or plants or to bore holes in the shells of other animals.

Reclamation. A process in which wasteland is converted into usable space. Inundated shorelines are often recovered by diking back the sea.

Red tide. A phenomenon produced by the explosive growth of certain dinoflagellates, causing the sea to turn red. Toxins produced by these animals can kill fish on a large scale and poison shellfish, making them unsafe for human consumption.

Reef. An underwater structure built by carbonate-secreting organisms like coral and encrusting algae. Fringing reefs are connected to shore and run parallel to the coast. Barrier reefs also parallel the coast, but they are separated from shore by a lagoon.

Refraction. The bending of a light ray or sound wave as it passes through one medium and then through another medium of different density. Ocean waves are also refracted or bent by submerged or shoreline features.

Regeneration. The natural replacement of lost tissue or an entire body part.

Reproduction. *See* Asexual reproduction; Sexual reproduction.

Rift. A fault in the earth's crust where the movement of material is away from the fault in a lateral direction. The line of submerged mountains that bisects the Atlantic Ocean in a north-south direction is often referred to as the Mid-Atlantic Rift.

"Ring of fire." Active volcano and earthquake activity around the margin of the Pacific Ocean, a result of large-scale movements of the earth's crust.

Rip Currents. Nearshore, wave-induced water movements that push water toward the beach. In the surf zone, the water moves parallel to the beach. When these longshore currents combine, they produce jetlike streams of water, a few tens of meters across, which move seaward through the surf. Such rip currents have swept many a bather out to sea.

S

Salinity. A measure of the amount of salt (usually determined by chloride ions) dissolved in water; expressed in parts per thousand.

Sarcopterygii. The lobe-finned, air-breathing fish.

Sargasso Sea. The region of the North Atlantic between the West Indies and the Azores, characterized by an abundance of the drifting seaweed *Sargassum* and by almost motionless water inside the circulating North Atlantic gyre.

Scyphozoan. A coelenterate which features a well-developed medusoid stage. Jellyfish are examples of this class, in essence a swimming, upside-down anemone.

Sediment. A particle or bed of particles that is the result of the weathering and erosion of preexisting rocks or organic materials. When the sediments become lithified (compressed or cemented into rock), they become *sedimentary rocks.*

Seine. A type of fishing net that hangs vertically in the water, its bottom edge weighted and its upper edge supported by floats. Seines trap fish by enclosing them in a confined area. Varieties are purse seines and haul seines. Other fishing apparatus include *long lines* (with many hooks), *trawls* (which are dragged along behind a ship), and *gill nets* (a wall of netting that hangs suspended in the water, trapping fish within its mesh).

Sial. The rocks, rich in aluminum, sodium, and potassium, that compose the upper layers of the earth's crust.

Silicoflagellate. Flagellated, planktonic organisms with siliceous shells; common in most of the colder parts of the ocean.

Sirenians. A group of herbivorous marine mammals that includes the manatee and dugong; distantly related to the land elephant.

Sima. The rocks of the lower part of the earth's crust, which are rich in magnesium, calcium, and iron minerals.

Sonar. A system of echolocation used to detect objects underwater; acronym for SOund NAvigation Ranging. Applicable to man's and dolphin's systems.

Spawn. The eggs of aquatic animals, including fish, invertebrates, and amphibians. The larval stage of many bivalves is referred to as *spat.*

Species. The most precise unit of taxonomy; a group of organisms that can interbreed with members of no other group.

Spectrum. The ordered arrangements of light and radiant energy, usually classified by wavelengths. The device to measure this distribution is called a *spectrophotometer,* and produces a *spectrogram.*

Spermatozoa. The male gamete in a sexually reproducing animal; usually a tadpole-shaped motile structure with a long whiplike tail.

Spicule. A needlelike supporting structure, which may be composed of calcium carbonate, silica, or chitin; characteristically found in sponges.

Statocyst. A balance organ found in many organisms; also called an *otocyst.* It consists of a fluid-filled sac and may contain a *statolith* or otolith (a small concretion that moves, under the influence of gravity, with the fluid and tells the animal which end is up or down).

Stomiatoid. Any of a group of bathypelagic fish commonly called viperfish or dragonfish, which may possess fanglike teeth and luminous barbels.

Storm surge. A change in sea level associated with a storm. They occasionally cause catastrophes in low-lying areas.

Stratosphere. The portion of the earth's atmosphere that is relatively uniform in temperature, above the *troposphere* and below the *mesophere,* beginning about seven miles up.

The upper boundary of the troposphere reaches about 5 miles above the poles and 10 miles above the equator. The stratosphere extends 10 to 15 miles higher, and the mesosphere goes up to the 50-mile level. The next region is the *ionosphere,* which extends to an altitude of 350 to 600 miles. The outermost atmospheric layer is the *exosphere*—900 miles of thinly dispersed helium surrounded by a hydrogen layer that extends 4000 miles before it tapers off into the void of space. Atoms and molecules in the exosphere are so far apart they seldom collide and some are lost forever to space.

Stridulate. To produce by friction rapid vibrations sounding like chirping; frequently done by many marine invertebrates and some fish and mammals.

Stroboscope. A device to make moving objects appear stationary by emitting bright light impulses at variable intervals. Stroboscopy is used to analyse movements and to measure revolutions per minute.

Stromatolite. A mat of algae and sediment which may harden and fossilize. Some stromatolites are three billion years old.

Sublittoral zone. A division of the benthic province extending from approximately a depth of 150 feet (50 meters) to the edge of the continental shelf, a depth of about 600 feet (200 meters).

Submersible. A controlled, free-diving underwater vessel that does not have ballast carried externally (as in a submarine). Generally submersibles are small and used for research.

Surf. Waves breaking in a coastal area.

Swell. Waves that have traveled away from the site of generation, usually in a regular pattern.

Swim bladder. A gas-filled sac found in many fish, which helps them maintain buoyancy and sometimes aids them in respiration. For hearing and sound production it may act as a resonating chamber.

Symbiosis. Any relationship between two or more organisms. Sometimes confused with mutualism. Encompasses the subdivisions of commensalism, parasitism, and mutualism.

Syncline. A folded layer or layers of rock with the center depressed relative to the sides and with the youngest rocks exposed in the center after erosion levels the surface.

Synergism. In chemistry, the process by which two or more substances reinforce each other's effects, or in combination produce a new effect. Many poisons show this property.

Synthesis. The creation of a substance, usually by an organism or experimental process.

T

Taxonomy. That branch of science dealing with the systematic classification of living things.

Tectonics. Movements and mountain-building forces within the earth; also the study of those forces and the changes they produce.

Teleostei. A superorder within the class Osteichthys (bony fishes) to which modern bony fish belong. Living bony fish *not* belonging to this group include the sturgeon, paddlefish, garpike, and bowfin.

Thermistor. A solid-state electronic device used to measure water temperature. The thermistor's electrical resistance decreases as the water temperature increases. This change in resistance is monitored and converted into temperature differences.

Thermal inversion. A atmospheric phenomenon that occurs when a warm air mass overrides a cooler one. Pollutants from below are trapped beneath the warm air in a horizontal layer.

Thermocline. A layer of water between warm surface water and cool deep water where temperature decreases rapidly with depth. A thermocline can act as a barrier to the transmission of sound in the sea, as well as to vertical migrations of fish.

Thermohaline circulation. Movement of water caused by differences in salinity and temperature. In the antarctic region, for example, cold salty water, which has a high density, sinks and flows northward along the bottom to the equator.

Thermoregulation. Maintenance of body temperature at a specific level by means of heat production and various methods designed to conserve heat. Thermoregulation is characteristic of warm-blooded animals and promotes homeostasis.

Tide pool. A pocket of seawater isolated by the receding tide. It is an unusual environment; salinity and temperature can fluctuate greatly before the next tide comes in.

Till. Sediment carried or deposited by a glacier. *Tillite* is a sedimentary rock composed of this.

Topography. The surface features of an area, including relief, lakes, rivers, and so on. Oceanographers are interested in the topography of the ocean bottom and spend a great deal of time charting the depths of the sea.

Trade winds. Persistent winds that blow from about 30° N. latitude toward the equator from the northeast, and from 30° S. latitude toward the equator, from the southeast. These winds drive the North and South Equatorial Currents in a westward direction and indirectly the Equatorial Countercurrent in an eastward direction.

Trawl. A specific type of oceanographic sampling and commercial fishing gear, most often towed behind a moving ship. A trawl can be moved over the bottom, where it collects benthic animals, or at a midwater depth, where it can sample fish life with some success.

Trenches. Deep fissures in the ocean floor where old crust is destroyed and forced back into the interior of the earth.

Trilobite. Any of a primitive group of arthropods that became extinct about 230 million years ago.

Tsunami. A seismic sea wave produced by a sudden movement of the ocean bottom. It can race undetected across the ocean at hundreds of miles per hour, but when it reaches shallower shore areas, it can grow to heights of 100 feet or more, wreaking havoc on these far-distant shores.

Tundra. A vast, nearly level, treeless plain characteristic of the arctic region.

Turbidity current. A dense, sediment-laden current of water that usually flows downward through less dense water along the slope of a continent. Some submarine canyons may have been cut by turbidity currents. Few have actually been observed; their dynamics and effects on submarine erosion are poorly understood.

U

Ultrasonic frequencies. Mechanical vibrations (sound) above the range of human detectability, which is 16 to 16,000 cycles per second. Whales and dolphins have hearing particularly sensitive to ultrasonic sound.

Undertow. A current of water moving seaward under the breaking surf. More generally, it can be any current of water moving beneath the surface water in a different direction.

Upwelling. An oceanic phenomenon wherein deep water is drawn to the surface. Upwelling can be induced by wind and surface currents. Bottom waters are rich in nutrients, and areas of upwelling are areas of high biological productivity.

V

Vacuole. A relatively clear, fluid-filled cavity within a cell. Vacuoles serve a variety of functions, from discharge of water and waste products to food and enzyme storage.

Van Allen Belt. A belt of intense ionizing radiation that surrounds the earth in the outer reaches of the atmosphere. It is formed by charged particles emanating from the sun and outer space which are attracted and trapped by the earth's magnetic field.

Vertebrate. Any animal possessing a backbone and a cranium. Vertebrata, the large subphylum of chordate animals, includes all mammals, birds, reptiles, amphibians, and fish.

Virus. Ultramicroscopic particles considered by some to be alive and by others to be complex proteins that sometimes include nucleic acids and enzymes. Viruses can multiply only in connection with a living cell.

Viscosity. The internal friction of a fluid, caused by attractions of the molecules for each other, which makes it resistant to flow.

Viviparous. The retention of the egg within the mother during development and subsequent live birth, seen in mammals. In contrast to *oviparous,* in which the egg is laid and development takes place externally, seen in birds and some reptiles, and *ovoviviparous,* in which development is internal but no placental connection ever forms between mother and embryo, seen in sharks and other fish.

W

Wavelength. If one pictures ripples emanating from a pebble dropped in a pond, the wavelength is the distance between one wave crest and the next; or between one wave trough and the next; or between any point on one wave and the corresponding point on the next one. For sound, a wavelength is the distance between one zone of compression or refraction and another.

X

Xiphosura. A nearly extinct subclass of arthropods whose fossil record dates back to the Ordovician period, 450 million years ago. Three genera and five species of Xiphosurans exist today. The most common representative is *Limulus,* the horseshoe crab, a true living fossil.

Y

Yolk sac. A sac containing yolk, a nutrient for developing embryos that consists mainly of fat and protein. Commonly seen in fish embryos.

Z

Zooplankton. Tiny animals, rarely more than a few millimeters in length, that are unable to counter the movements of water in the oceans and are carried away, drifting at the mercy of the current.

Zooxanthellae. Symbiotic algae that live in the tissue of most reef coral, as well as other species. These dinoflagellate relatives require sunlight for photosynthesis, a factor that may restrict reefs to sunlit waters. The coral animals utilize the oxygen that the algae produce.

Zygote. The fertilized ovum in plants and animals; more technically, it is the diploid cell resulting from the fusion of male and female gametes.

Bibliography

General Ocean Science

Carson, Rachel L. *The Sea Around Us.* 2nd ed., rev. New York: Oxford University Press, 1961.

———. *The Edge of the Sea.* Boston: Houghton Mifflin, 1955.

Coker, R. E. *This Great and Wide Sea.* New York: Harper & Row, Harper Torchbooks, 1962.

Cowen, Robert C. *Frontiers of the Sea.* Garden City: Doubleday, 1960.

Cromie, William J. *Exploring the Secrets of the Sea.* Englewood Cliffs: Prentice-Hall, 1962.

Deacon, G. E. R. *Oceans.* London: Hamlyn, 1968.

Dietrich, Gunther. *General Oceanography: An Introduction.* Translated by Deodor Ostapoff. New York: Interscience, 1963.

Dugan, James, et al. *World Beneath the Sea.* Washington, D.C.: National Geographic Society, 1967.

Engel, Leonard, and the Editors of *Life* magazine. *The Sea.* New York: Time Inc., 1961.

Fairbridge, Rhodes W., ed. *The Encyclopedia of Oceanography.* New York: Reinhold, 1966.

Firth, Frank E., ed. *The Encyclopedia of Marine Resources.* New York: Van Nostrand, Reinhold, 1969.

Heberlein, Hermann. *Le monde sous-marin.* Zurich: BEA, 1959.

Herring, Peter J., and Malcolm R. Clarke. *Deep Oceans.* London: Arthur Barker, 1971.

Laurie, Alec. *The Living Oceans.* London: Aldus Books, 1972.

Long, Edward J. *New Worlds of Oceanography.* New York: Pyramid Publications, 1965.

Olschki, Alessandro, ed. *SUB, enciclopedia del subacqueo.* Florence: Sadea/Sansoni, 1968.

Outhwaite, Leonard. *The Ocean.* London: Constable, 1961.

Ross, David A. *Introduction to Oceanography.* New York: Appleton-Century-Crofts, 1970.

Scientific American. *Oceanography.* San Francisco: W. H. Freeman, 1971.

Sverdrup, H. U., Martin W. Johnson, and Richard H. Fleming. *The Oceans, Their Physics, Chemistry and General Biology.* New York: Prentice-Hall, 1942.

Weyl, Peter K. *Oceanography.* New York: Wiley, 1970.

The Poles and the Abyss

Bruun, Anton Frederic, et al. *The Galathea Deep Sea Expedition.* Translated by Reginald Spink. London: Allen & Unwin, 1956.

Heezen, Bruce C., and Charles D. Hollister. *The Face of the Deep.* New York and London: Oxford University Press, 1971.

Holdgate, M. W., ed. *Antarctic Ecology.* London and New York: Academic Press, 1970.

Idyll, C. P. *Abyss, the Deep Sea and the Creatures That Live in It.* New York: Thomas Y. Crowell, 1964.

King, H. G. R. *The Antarctic.* New York: Arco, 1969.

Menzies, Robert J., Robert Y. George, and Gilbert T. Rowe. *Abyssal Environment and Ecology of the World Oceans.* New York: Wiley, 1973.

Stefansson, Vilhjalmur. *Hunters of the Great North.* New York: Harcourt, Brace, 1922.

Earth and Ocean Sciences

Badgley, Peter C., Leatha Miloy, and L. Childs, eds. *Oceans from Space.* Houston: Gulf Publishing, 1969.

Bascom, Willard. *Waves and Beaches.* Garden City: Doubleday, 1964.

Bouteloup, Jacques. *Vagues, marées, courants marins.* Paris: Presses Universitaires de France, 1968.

Chapin, Henry, and F. G. Walton Smith. *The Ocean River.* New York: Scribner, 1962.

Kummel, Bernhard. *History of the Earth.* 2nd ed. San Francisco: W. H. Freeman, 1970.

Starfish

Emperor Angelfish

Alcyonacean

Observation Chamber of *Calyps*

Lacombe, Henri. *Les energies de la mer.* Paris: Presses Universitaires de France, 1968.
———. *Les mouvements de la mer: courants, vagues et houle, marées.* Paris: Doin, 1971.
Leet, L. Don, and Sheldon Judson. *Physical Geology.* Englewood Cliffs: Prentice-Hall, 1971.
Menard, H. W. *Marine Geology of the Pacific.* New York: McGraw-Hill, 1964.
Pettersson, Hans. *The Ocean Floor.* New Haven: Yale University Press, 1954.
Shepard, Francis P. *The Earth Beneath the Sea.* New York: Atheneum, 1964.
Shepard, Francis P., and Harold R. Wanless. *Our Changing Coastlines.* New York: McGraw-Hill, 1971.
Smith, F. G. Walton. *The Seas in Motion.* New York: Thomas Y. Crowell, 1973.
Trewartha. Glenn T. *An Introduction to Climate.* 4th ed. New York: McGraw-Hill, 1968.

Evolution

Beerbower, James R. *Search for the Past.* 2nd ed. Englewood Cliffs: Prenctice-Hall, 1968.
Buettner-Janusch, John. *Origins of Man.* New York, London, and Sydney: John Wiley, 1966.
Colbert, Edwin H. *Evolution of the Vertebrates.* 2nd ed. New York: John Wiley, 1955.
Darwin, Charles E. *The Origin of Species.* New York: Random House, The Modern Library, 1936.
Romer, Alfred S. *The Vertebrate Story.* 4th ed. Chicago and London: University of Chicago Press, 1959.

Marine Archaeology

Bass, George F. *Archaeology Underwater.* New York: Praeger, 1966.
Bass, George F., ed. *A History of Seafaring.* New York: Walker, 1973.
Cleator, P. E. *Underwater Archaeology.* New York: St. Martin's Press, 1973.
Dumas, Frédéric. *Deep Water Archeology.* Translated by Honor Frost. Chester Springs, Pa.: DuFour Editions, 1962.
Marx, Robert F. *The Lure of Sunken Treasure.* New York: David McKay, 1973.
———. *Port Royal Rediscovered.* Garden City: Doubleday, 1973.
Peterson, Mendel. *History Under the Sea.* Washington, D.C.: Smithsonian Institution, 1954.
Potter, John, Jr. *The Treasure Diver's Guide.* Garden City: Doubleday, 1972.
Throckmorton, Peter. *Shipwrecks and Archaeology, The Unharvested Sea.* Boston: Little, Brown, 1969.
———. *The Lost Ships.* London: Jonathan Cape, 1965.

Man and the Sea

Balder, A. P. *Complete Manual of Skin Diving.* New York: Macmillan, 1968.
Barnaby, K. C. *Some Ship Disasters and Their Causes.* Maritime Library Series. Cranbury, N.J.: A. S. Barnes, 1970.
Beebe, William. *Half Mile Down.* New York: Duell, Sloan and Pierce, 1955.
Clark, Eugenie. *Lady With a Spear.* New York: Harper Brothers, 1953.
———. *The Lady and the Sharks.* New York: Harper & Row, 1969.
Cousteau, Jacques-Yves, and James Dugan. *The Living Sea.* New York: Harper & Row, 1963.
Cousteau, Jacques-Yves, and James Dugan, eds. *Captain Cousteau's Underwater Treasury.* New York: Harper & Row, 1959.
Cousteau, Jacques-Yves, and Frederic Dumas. *The Silent World.* New York: Harper & Row, 1953.
Darwin, Charles. *The Voyage of the Beagle.* New York: Bantam Books, 1972.
Davis, Robert H. *Deep Diving and Submarine Operations.* 7th ed. Chessington, Surrey: Siebe, Gorman and Co., 1962.

Star II—Submersible

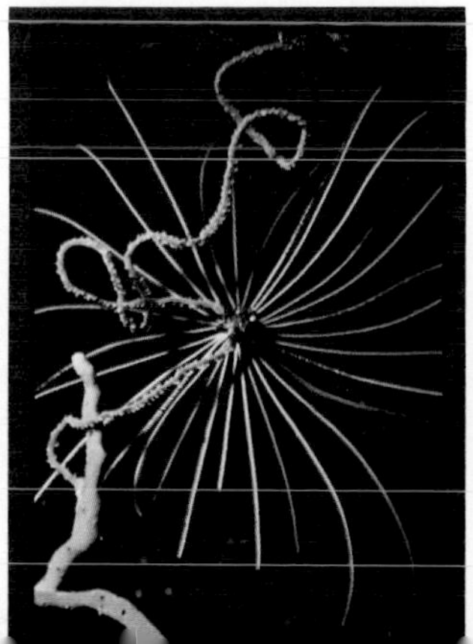
Crinoid

Alaskan Fur Seals

Calypso Diver and Octopus

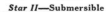

Deacon, G. E. R., ed. *Seas, Maps, and Men.* London: Crescent Books, 1962.

Dugan, James. *Man Under the Sea.* New York: Collier Books, 1965.

Dugan, James, and Richard Vahan, eds. *Men Under Water.* Philadelphia: Chilton Books, 1965.

Frey, Hank, and Shaney Frey. *130 Feet Down, A Handbook for Hydronauts.* New York: Harcourt, Brace and World, 1961.

Gaskell, T. F. *Under the Deep Oceans: Twentieth-Century Voyages of Discovery.* New York: Norton, 1960.

Gilbert, Perry W., ed. *Sharks and Survival.* Boston: Heath, 1963.

Gordon, Bernard L., ed. *Man and the Sea.* Garden City: Doubleday, Natural History Press, 1972.

Groueff, Stéphane. *L'homme et la mer.* Paris: Larousse, Paris-Match, 1973.

Groupe d'Etudes et de Recherches Sous-Marine. *La Plongée.* Paris: B. Arthaud, 1955. (Published in English under title *Complete Manual of Free Diving* [New York: Putnam, 1957])

Houot, Georges, and Pierre Willm. *2000 Fathoms Down.* New York: Dutton, 1955.

Lee, Owen. *Skin Diver's Bible.* Garden City: Doubleday, 1968.

Miles, S. *Underwater Medicine.* Philadelphia: Lippincott, 1966.

Penzias, Walter, and M. W. Goodman. *Man Beneath the Sea.* New York, London, Sydney, and Toronto: Wiley-Interscience, 1973.

Piccard, Jacques, and Robert S. Dietz. *Seven Miles Down: The Story of the Bathyscaphe Trieste.* New York: Putnam, 1961.

Schlee, Susan. *The Edge of an Unfamiliar World, A History of Oceanography.* New York: Dutton, 1973.

Shenton, E. H. *Exploring the Ocean Depths.* New York: Norton, 1968.

Smith, William D. *Northwest Passage.* New York: American Heritage Press, 1970.

Soule, Gardner. *Undersea Frontiers.* New York: Rand McNally, 1968.

Stenuit, Robert. *The Deepest Days.* New York: Coward-McCann, 1966.

Stephens, William M. *Science Beneath the Sea.* New York: Putnam, 1966.

Stewart, Harris B. *Deep Challenge.* Princeton: Van Nostrand, 1966.

United States Department of the Navy. *U.S. Navy Diving Manual.* Washington, D.C.: U.S. Government Printing Office, 1963.

Vaissiere, Raymond. *L'Homme et le monde sous-marin.* Paris: Larousse, 1969.

Villiers, Alan, et al. *Men, Ships, and the Sea.* Washington, D.C.: National Geographic Society, 1963.

Mammals

Andersen, Harald T., ed. *The Biology of Marine Mammals.* New York and London: Academic Press, 1969.

Burton, Robert. *The Life and Death of Whales.* New York: Universe Books, 1973.

Howell, A. Brazier. *Aquatic Mammals.* New York: Dover, 1970.

Maxwell, Gavin. *Seals of the World.* Boston: Houghton-Mifflin, 1967.

Norris, Kenneth S., ed. *Whales, Dolphins and Porpoises.* Berkeley: University of California Press, 1966.

Perry, Richard. *The World of the Polar Bear.* Seattle: University of Washington Press, 1966.

Ridgeway, Sam H., ed. *Mammals of the Sea.* Springfield, Ill.: Charles C. Thomas, 1972.

Slijper, E. J. *Whales.* New York: Basic Books, 1962.

Small, George L. *The Blue Whale.* New York and London: Columbia University Press, 1971.

Undersea Resources

Bardach, John. *Harvest of the Sea.* London: Allen & Unwin, 1969.

Brandt, Andres von. *Fish Catching Methods of the World.* 2nd ed., rev. London: Fishing News, 1972.

Gulland, J. A., ed. *The Fish Resources of the Ocean.* London: Fishing News, 1971.

Idyll, C. P. *The Sea Against Hunger.* New York: Thomas Y. Crowell, 1970.

Iverson, E. S. *Farming the Edge of the Sea.* London: Fishing News, 1968.

McKee, Alexander. *Farming the Sea,* New York: Thomas Y. Crowell, 1969.

Mero, John L. *The Mineral Resources of the Sea.* New York: American Elsevier, 1964.

Milne, P. H. *Fish and Shellfish Farming in Coastal Waters.* London: Fishing News, 1972.

Walford, Lionel A. *Living Resources of the Sea.* New York: Ronald, 1958.

Natural History

Abbott, R. Tucker. *Sea Shells of the World: A Guide to the Better-Known Species.* Edited by H. S. Zim. New York: Golden Press, 1962.

Alexander, W. B. *Birds of the Ocean.* New York: Putnam, 1954.

Barnes, Robert D. *Invertebrate Zoology.* Philadelphia and London: Saunders, 1963.

Boney, A. D. *A Biology of Marine Algae.* London: Hutchinson Educational, 1966.

Buchsbaum, Ralph. *Animals Without Backbones.* Chicago and London: University of Chicago Press, 1965.

Buchsbaum, Ralph, and Lorus J. Milne. *The Lower Animals: Living Invertebrates of the World.* Garden City: Doubleday, 1967.

Bustard, Robert. *Sea Turtles.* New York: Taplinger, 1972.

Carr, Archie. *So Excellent a Fishe: A Natural History of Sea Turtles.* Garden City: Doubleday, Natural History Press, 1967.

———. *The Windward Road.* New York: Knopf, 1955.

Curtis, Brian. *The Life Story of the Fish.* New York: Dover, 1961.

Eisenberg, J. F., and William S. Dillon, eds. *Man and Beast: Comparative Social Behavior.* Washington, D.C.: Smithsonian Institution, 1971.

Eibl-Eibesfeldt, Irenaus. *Ethology, The Biology of Behavior.* New York: Holt, Rinehart, Winston, 1970.

Halstead, B. W. *Poisonous and Venomous Marine Animals of the World.* 3 vols. Washington, D.C.: U.S. Government Printing Office, 1965.

Hardy, A. C. *The Open Sea: Its Natural History—Fish and Fisheries.* London: Collins, 1959.

———. *The Open Sea: Its Natural History—The World of Plankton.* London: Collins, 1956; Boston, Houghton Mifflin, 1959.

Herald, Earl S. *Living Fishes of the World.* Garden City: Doubleday, 1971.

Hickman, Cleveland P. *Integrated Principles of Zoology.* St. Louis: Mosby, 1961.

Hyman, L. M. *Invertebrates.* 5 vols. New York: McGraw-Hill, 1940—1959.

Johnson, M. E., and Harry James Snook. *Seashore Animals of the Pacific.* New York: Dover, 1967.

Lagler, Karl F., John E. Bardach, and Robert R. Miller. *Ichthyology.* Ann Arbor: University of Michigan Press, 1962.

Lane, Frank W. *The Kingdom of the Octopus: The Life History of the Cephalopoda.* New York: Sheridan, 1960.

MacGinitie, G. E., and Nettie MacGinitie. *Natural History of Marine Animals.* New York: McGraw-Hill, 1968.

Marshall, N. B. *The Life of Fishes.* New York: Universe Books, 1966.

———. *Explorations in the Life of Fishes.* Cambridge, Mass.: Harvard University Press, 1971.

Newell, G. E., and R. C. Newell. *Marine Plankton: A Practical Guide.* London: Hutchinson Education, 1963.

Odum, Eugene P. *Fundamentals of Ecology.* Philadelphia: Saunders, 1971.

Ommanney, F. D., and the Editors of *Life* magazine. *The Fishes.* New York: Time Inc., 1963.

Orr, Robert T. *Animals in Migration.* New York: Macmillan, 1970.

Peres, Jean-Marie. *La Vie dans les mers.* Paris: Presses Universitaires de France, 1972.

———. *Clefs pour l'oceanographie.* Paris: Seghers, 1972.

Randall, John E. *Caribbean Reef Fishes.* Neptune City, N.J.: T. F. H. Publishers, 1968.

Ricketts, Edward F., and Jack Calvin. *Between Pacific Tides.* 3rd ed. Stanford: Stanford University Press, 1952.

Schmitt, Waldo L. *Crustaceans.* Ann Arbor: University of Michigan Press, 1965.

Schultz, Leonard P., et al. *Wondrous World of Fishes.* Washington, D.C.: National Geographic Society, 1965.

Smith, J. L. B. *The Search Beneath the Sea.* New York: Henry Holt, 1956.

Torchio, Menico. *La vita nel mare.* Novara: Istituto geografico de Agostini, 1972.

Wilson, D. P. *Life of the Shore and Shallow Sea.* London: Nicholson and Watson, 1951.

Conservation

Carr, Donald E. *Death of the Sweet Waters.* New York: Norton, 1966.

Carson, Rachel. *Silent Spring.* Boston: Houghton Mifflin, 1962.

Commoner, Barry. *The Closing Circle.* New York: Knopf, 1971.

Ehrlich, Paul R., and Anne M. Ehrlich. *Population, Resources, Environment.* 2nd ed. San Francisco: W. H. Freeman, 1972.

Kay, David A., and Eugene B. Skolnikoff. *World Eco-Crisis.* Madison: University of Wisconsin Press, 1972.

Marx, Wesley. *The Frail Ocean.* New York: Ballantine Books, 1967.

Moorcraft, Colin. *Must the Seas Die?* Boston: Gambit, 1973.

Index

This index is a complete reference guide to Volumes I through XIX. Many subjects are discussed, illustrated, or shown in photographs in a number of volumes, and with this tool the reader can easily locate all the material available on a particular topic. The list of volumes below includes a general statement of each one's content, and the sample entry indicates the elements that appear throughout the index.

Sample Index Entry

Algae, I 32; II 108; IV **47**
  antarctic, XVI 103
  See also Amphiroa

Main entry: Algae
Subentry: antarctic
Volume number: Indicated by roman numeral
Illustration or photograph: Indicated by bold face numeral
Cross-reference: Indicated by *See* or *See also*

I. Oasis in Space—The earth, its water, and its marine life.
II. The Act of Life—Reproduction of life in the sea.
III. Quest for Food—Location, capture, and uses of food.
IV. Window in the Sea—Light and vision underwater.
V. The Art of Motion—Propulsion under and on the sea.
VI. Attack and Defense—Conflict and survival in the sea.
VII. Invisible Messages—Sound—scent—electricity as media.
VIII. Instinct and Intelligence—Basic drives and behavior.
IX. Pharaohs of the Sea—Coral polyps: the reef builders.
X. Mammals in the Sea—The sea's warm-blooded creatures.
XI. Provinces of the Sea—Earth dynamics and ocean regions.
XII. Man Re-enters the Sea—Submarines and diving.
XIII. A Sea of Legends—Myths and inspiration of the sea.
XIV. The Adventure of Life—Evolution of life in the sea.
XV. Outer and Inner Space—Man's study of the oceans.
XVI. The White Caps—Life and conditions at the poles.
XVII. Riches of the Sea—Its food, minerals, and recreation.
XVIII. Challenges of the Sea—Facing the elements.
XIX. The Sea in Danger—Man's pollution of the oceans.
XX. Guide to the Sea and Index

A

Colonial Tunicates

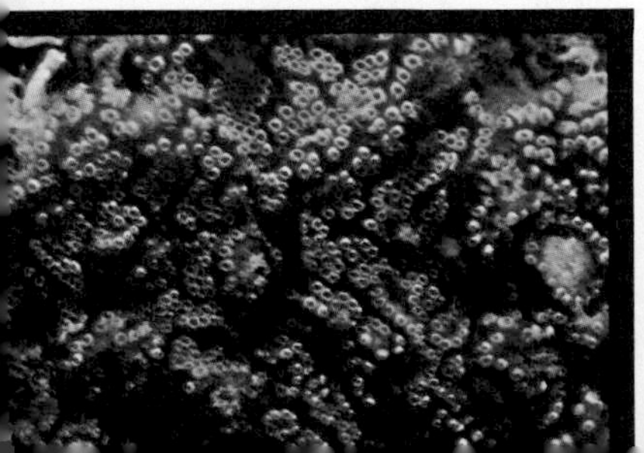

Diver Collecting Red Coral

Alcyonacean

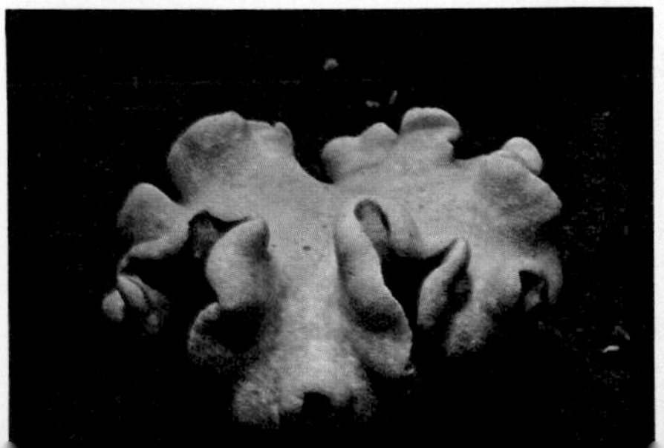

Diver and Dolphin

Calypso

Red Sea Reef Fish

Galápagos Iguanas

Antarctic

D

I

J

K

L

Q

R

S

(Vol. **XX**)

ILLUSTRATIONS AND CHARTS:

Howard Koslow—92-93; S. Harold Reuter, M.D.—40-41, 42.

PHOTO CREDITS:

Bruce Coleman, Inc.: Jeff Foott—79 (bottom), Oxford Scientific Films—5, 75 (top), L. L. T. Rhodes—22-23 (top), R. & J. Spurr—48-49; W. E. Ferguson—82 (top); Freelance Photographers Guild: Bob Gladden—80 (third from top), Jerry Jones—99 (bottom right), C. G. Maxwell—76 (third from top); Hughes Aircraft Company (from transparency supplied by NASA)—34; Hyperion Sewage Plant—74; Jack McKenney—76 (bottom), 78 (third from top), 104; Richard C. Murphy—26, 75 (middle), 76 (top; second from top), 78 (second from top), 78 (bottom), 79 (second from top), 80 (top), 99 (top left), 102 (top); NASA—35; Naval Undersea Center, San Diego—16 (bottom); Photography Unlimited: Ron Church—9; Dr. David Schwimmer—25, 50-51, 79 (third from top), 81 (third from top), 81 (bottom), 83 (bottom); The Sea Library: Jim and Cathy Church—33, 75 (bottom), Jack Drafahl—98 (top right), 99 (top right), 99 (bottom left), Chuck Eilers—22 (bottom), Robert B. Evans—102 (bottom), Daniel W. Gotshall—82 (bottom), Hyperion Sewage Plant—58, S. Keiser—89, Carl Roessler—77 (bottom), 78 (top), 81 (second from top), 101 (left), The Sea Library—79 (top), Valerie Taylor—98 (top left; bottom), U.S. Department of Fish and Game—83 (second from top), S. Williams—27; E. A. Shinn—77 (top; middle), 103 (left); Tom Stack & Associates: Neville Coleman, AMPI—103 (right), Fred Livingston, Jr.—28; Taurus Photos: Charlene Burch—37, Dick Clarke—61, Anthony Merciera—82 (second from top), Dave Woodward—47, 67, 81 (top), 101 (right); Myron Wang—55, 82 (third from top).